LEARNING AND TEACHING NUMBER THEORY

Recent Titles in
Mathematics, Learning, and Cognition
Monograph Series of the *Journal of Mathematical Behavior*
Carolyn A. Maher and Robert Speiser, Series Editors

Five Women Build a Number System
Robert Speiser and Chuck Walter

LEARNING AND TEACHING NUMBER THEORY

Research in Cognition and Instruction

Edited by
Stephen R. Campbell and Rina Zazkis

Mathematics, Learning, and Cognition
Monograph Series of the *Journal of Mathematical Behavior*, Volume 2
Carolyn A. Maher and Robert Speiser, Series Editors

Ablex Publishing
Westport, Connecticut • London

Library of Congress Cataloging-in-Publication Data

Learning and teaching number theory : research in cognition and instruction / edited by Stephen R. Campbell and Rina Zazkis.

p. cm.—(Mathematics, learning, and cognition)

Includes bibliographical references and index.

ISBN 1–56750–652–6 (alk. paper)—ISBN 1–56750–653–4 (pbk. : alk. paper)

1. Number theory—Study and teaching. I. Campbell, Stephen R. II. Zazkis, Rina. III. Series.

QA241.L43 2002

510 s—dc21

[512'.7071] 2001031649

British Library Cataloguing in Publication Data is available.

Library of Congress Catalog Card Number: 2001031649

ISBN: 1–56750–652–6

1–56750–653–4 (pbk.)

First published in 2002

Ablex Publishing, 88 Post Road West, Westport, CT 06881

An imprint of Greenwood Publishing Group, Inc.

www.ablexbooks.com

Printed in the United States of America

The paper used in this book complies with the Permanent Paper Standard issued by the National Information Standards Organization (Z39.48–1984).

10 9 8 7 6 5 4 3 2 1

Contents

1 Toward Number Theory as a Conceptual Field 1
Stephen R. Campbell and Rina Zazkis

2 Coming to Terms with Division: Preservice Teachers' Understanding 15
Stephen R. Campbell

3 Conceptions of Divisibility: Success and Understanding 41
Anne Brown, Karen Thomas, and Georgia Tolias

4 Language of Number Theory: Metaphor and Rigor 83
Rina Zazkis

5 Understanding Elementary Number Theory at the Undergraduate Level: A Semiotic Approach 97
Pier Luigi Ferrari

6 Integrating Content and Process in Classroom Mathematics 117
Anne R. Teppo

7 Patterns of Thought and Prime Factorization 131
Anne Brown

8 What Do Students Do with Conjectures? Preservice Teachers' Generalizations on a Number Theory Task 139
Laurie D. Edwards and Rina Zazkis

9 Generic Proofs in Number Theory 157
Tim Rowland

10 The Development of Mathematical Induction as a Proof Scheme: A Model for DNR-Based Instruction 185
Guershon Harel

11 Reflections on Mathematics Education Research Questions in Elementary Number Theory 213
Annie Selden and John Selden

Author Index 231

Subject Index 235

About the Contributors 243

1

Toward Number Theory as a Conceptual Field

Stephen R. Campbell and Rina Zazkis

Since the beginning of the ancient Greek Pythagorean tradition over two and a half millennia ago, striving for a conceptual understanding of numbers and their properties, patterns, structures, and forms has constituted the heart, if not the soul, of mathematics and mathematical thinking. Today, a constellation of activities involving various operations, procedures, functions, relations, and applications associated with numbers occupy the main bulk of the K–12 mathematics curriculum. These activities are typically conducted under the auspices of arithmetic and algebra in various guises, such as counting and measuring with numbers, tabulating and graphing collections of numbers, and formulating and solving equations applied in less formal and more familiar "day-to-day," "real-world," situational contexts.

With all of the attention that has been given to informal meanings and familiar contexts in mathematics education these days, it seems that not too much consideration has been given to formal contexts concerned with properties and structures of number per se. Mathematical meaning is not just a matter of grounding concepts in familiar day-to-day real-world experiences. It is also a matter of developing the conceptual foundations for making clear and general abstract distinctions. A case in point has to do with understanding differences between whole numbers and rational numbers and the different kinds of procedures involved in operating with them. It is well known that learners have many procedural and semantic difficulties in this regard (e.g., Durkin & Shire, 1991; Greer, 1987; Mack, 1995; Silver, 1992). Teachers need to have and learners need to develop better structural understandings of these kinds of fundamental mathematical ideas (Kieran, 1992; Ma, 1999). Perhaps a more explicit and focused emphasis on number theory can help.

Carl Friedrich Gauss, the "prince of mathematicians," purportedly considered number theory, or "higher arithmetic," as the "queen of mathematics." Gauss, were he alive today, would likely be surprised, if not dismayed, to hear that despite the heavy emphasis on numbers and operations with numbers, *number theory* is far from enthroned within the curriculum. Nor could Gauss be told that very much research in mathematics education has been conducted with respect to learners' appreciation and understanding of her many subtle and profound ways.

Could it be that number theory still suffers from a time-honored perception that due to its austere abstract conceptual purity and theoretical nature, it has little or no practical value (Hardy, 1967)? Surely, any such view has been rendered indefensible today with the manifold applications of number theory with the advent of the information age (cf. Mathematical Sciences Education Board, 1989). The importance of number theory, and its perennial allure, is certainly not without practical applications.

And what is the role of number theory in the day-to-day practice of teaching mathematics? According to the 1989 National Council of Teachers of Mathematics (NCTM) *Curriculum and Evaluation Standards*: "Number theory offers many rich opportunities for explorations that are interesting, enjoyable, and useful. These explorations have payoffs in problem solving, in understanding and developing other mathematical concepts, in illustrating the beauty of mathematics, and in understanding the human aspects of the historical development of number" (p. 91). Despite these "rich opportunities," number theory is not typically considered a topic in and of itself in the K–12 curriculum. Given the central role of *number* in the mathematics curriculum and the central role of *the theory of numbers* in the history of mathematics, it is somewhat surprising that number theory does not play a more central role in the mathematics curriculum than it does. Perhaps more surprising, especially with regard to the cognitive development of mathematical understanding, is that such a paucity of attention has been given to research into learning and teaching number theory.

The main goal of this monograph is to identify and demonstrate some of the different kinds of problems and ways of thinking that can be investigated in a program of research into learning and teaching number theory and its implications for cognition and instruction. In so doing, we hope to begin charting out a few directions in which this work can progress. The balance of this chapter provides a brief review of number theory and its role in the K–16+ curriculum, along with an overview of the research in this volume and a general conceptual framework from which it can be considered.

NUMBER THEORY AS A SUBJECT AREA

Number theory is primarily concerned with the properties of the integers. As such, it constitutes a clear and well-defined subject content area. Introductory topics in number theory, such as divisibility or prime decomposition, are typically restricted to the domain of non-negative integers or whole numbers. More advanced studies in number theory bring algebraic and analytic notions—which incorporate nonintegral numerical domains and methods—to bear on the integers. Some studies in number theory also focus on certain properties of rational, real, and complex numbers that depend directly on the properties of the integers. Here, we characterize *elementary* number theory by restricting any particular means, that is, sequences, functions, equations, groups, and so on, used for studying the properties of integers to the domain of the integers. Methods for studying integers that implicate nonintegral domains will be taken to characterize topics in *advanced* number theory.

Number theory in the K–12 curriculum, as noted above, is not typically designated as a subject area of study in and of itself—at least not in the same sense that arithmetic, algebra, and geometry usually are. In fact, number theory is usually intertwined with these subjects. Some introductory topics of elementary number theory are occasionally given separate treatment in the K–12 curriculum and are, perhaps for that reason, explicitly included in professional development content and methods courses as well. These introductory topics include figurate numbers, whole number patterns and sequences, multiples, factors, divisors, primes, composites, prime decomposition, relatively prime numbers, divisibility, and divisibility rules.

At post-secondary levels, courses in number theory are offered as optional or required courses toward filling major or minor requirements in most undergraduate mathematics programs. Such courses usually take an axiomatic approach toward elementary number theory as formally defined over the ring of integers. These courses tend to include more rigorous and generalized treatments of topics such as prime, composite numbers, greatest common divisors, least common multiples, linear and polynomial congruence, ideals, quadratic residues and reciprocity, Diophantine equations, number theoretic functions, and so on. Continued fraction expansions of rational and irrational numbers, according to our characterization, would be an advanced topic that is often taught in these courses as well.

Also at post-secondary levels, various topics of elementary and advanced number theory are included in survey courses on abstract mathematics, and are often integrated into information and computer science courses on discrete mathematics, combinatorics, coding theory, and so on. This is a reflection of the growing importance of the many practical applications of number theory noted above. More advanced algebraic and analytical treatments of number theory are typically offered in graduate school programs in mathematics and computer science. The studies in this monograph focus mostly on introductory topics from elementary number theory as they are approached at the undergraduate and professional development levels.

NUMBER THEORY AND THE NCTM STANDARDS

Standard #6 for the middle grades in the NCTM (1989) *Curriculum and Evaluation Standards for School Mathematics* (henceforth referred to as *Standards*) is titled "number systems and number theory." It is noted therein that the central theme of this standard is with facilitating students' understanding of the underlying structure of mathematics and of arithmetic in particular. Number theory, however, has lost its status as a designated content standard in the new NCTM (2000) *Principles and Standards for School Mathematics* (henceforth referred to as *Principles and Standards*). The underlying attitude behind this shift seems to be that gaining proficiency with rational numbers should take priority and that students will, in any case, gain lots of experience in working with whole numbers in other problem-solving contexts (NCTM, 1998, p. 206). Consistent with this rationale, introductory topics in and relating to elementary number theory have largely been subsumed under content standards for "number and operation" and "algebra." Furthermore, the old *Standards* emphasis on the utility of number theory as helping learners to develop an understanding of the underlying structure of mathematics appears to be either absent or left implicit in the new *Principles and Standards*.

The new content standards in the *Principles and Standards*, however, are not necessarily intended to strictly designate specific subject content areas. Various topics of discrete mathematics (as a content standard for grades 9–12 in the *Standards)* and number theory (as a content standard for grades 5–8 in the *Standards*), now fall within other content standards. With regard to the former, it is noted that "the main topics of discrete mathematics are included, but they are distributed across the Standards, instead of receiving separate treatment, and they span the years from prekindergarten through grade 12" (NCTM, 2000, p. 31). Although not explicitly mentioned, similar qualifications seem warranted for number theory as well.

Whereas the emphasis on number theory for acquiring a deeper conceptual understanding of fundamental properties of numbers and numerical structures appears to have been left implicit in the *Principles and Standards*, its role with respect to problem solving and reasoning has been highlighted. In mathematics education research, studies in students' conceptions of proof have been drawing heavily on topics from number theory under the name of algebra (e.g., Healy & Hoyles, 2000). It is becoming more evident and widely recognized that number theory promotes mathematical reasoning skills in the secondary grades, particularly with respect to mathematical proof. This perception has not only *not* been diminished with the *Principles and Standards*; it has been extended downward into the middle grades (cf. NCTM 2000, pp. 264–265, p. 292, p. 345).

From numerical patterns (e.g., figurate numbers) and procedures (e.g., Euclidean algorithm) to concepts (e.g., divisibility) and reasoning (e.g., existence and uniqueness proofs), number theory offers a unified and substantial spectrum of dimensionality that can be called into service for the improvement of mathematics education. As such, number theory could readily stand alone as a content standard between "numbers and operations" and "algebra" ranging over the K–12 curriculum, having clear instantiations across all process standards. The extent to which it may fruitfully be considered and implemented as such is an important research question for investigating the learning and teaching of number theory.

EPISTEMOLOGICAL ORIENTATION

To one extent or another, all of the studies in this monograph fall within a constructivist tradition that views learners as interacting with the world while actively engaged with structuring and organizing their experience of interacting with that world. Contrary to popular misconceptions, constructivism does not entail a denial of an objective real world existing independently of our experience. There is, rather, simply no rational basis for affirming that we can come to know anything about that world that has not been structured in some way through our own individual and collective experience. A similar position, one may note, can readily be taken regarding the independent existence of mathematical realms.

Constructivism is part of an epistemological tradition dating back at least to Descartes's grounding of veritable truth in the human *cogito* and Kant's grounding of objective knowledge in the subjectivity of human cognition rather than in the world itself. Not constrained by its philosophical origins, constructivism, through the work of Piaget and others, has been recast in the scientific pursuits of genetic or natural epistemology—resulting in more developmental and operational approaches to investigating human cognition (e.g., Kornblith, 1994; Piaget, 1972).

In accord with Glasersfeld's (1982, 1995) interpretation of Piaget, constructivists have adopted a more pragmatic approach to truth and knowledge by abandoning notions that either can be legitimately considered absolute or infallible. According to this "radical constructivism," individuals, reflecting upon recurrent sensorimotor and perceptually guided experience, actively engage with their experiential world in the construction of procedures and concepts. Learners' abstract mathematical understandings are not assumed to be completely separable from or independent of concrete experience. Rather, such understandings are mediated by, and manifested through, the use of symbolic expressions, such as verbalized and written numerals, and activities involving concrete materials, such as manipulatives, diagrams, calculators, computer graphics, and so on. Practical utility and real-world applications aside, it is assumed that concrete sensorimotor experiences involved in articulating or manipulating mathematical expressions and materials form a necessary component in the development of any abstract understanding of mathematics.

Thus for constructivists, terms such as "abstract," "decontextualized," or "context-free" mathematical knowledge can be contentious and problematic if interpreted to mean existing independently of concrete sensorimotor experience or lived "real-world" contexts. Such interpretations presuppose or postulate ontological realms that exist independently of experience. A central tenant of constructivism, it bears repeating, is that there is no way of knowing anything about any metaphysical or mathematical realities independently of our own experience (see, e.g., Glasersfeld, 1995; Kant, 1787/1965). Be this as it may, there is a broad consensus that in order to understand and speak of objects in such realms at all, they must at least be thought, and spoken of, *as if* they do exist (cf. Dörfler, 1995; Glasersfeld, 1995; Kant, 1787/1965; Sfard, 1991).

A central theoretical concern of constructivism is to account for how human actions, experience, thought, and language develop and come to be structured in meaningful ways. A fundamental concept for constructivists engaged in investigating the broad spectrum of human cognition and experience is variously referred to as a scheme, schemata, or schema, after the Greek word *schema*—meaning appearance or shape. According to Vergnaud, "[a] scheme is defined as the invariant organization of action for a certain class of situations. This dynamic totality, introduced by Piaget (after Kant), to account for both 'sensory-motor skills' and 'intellectual skills' requires a strict and deep analysis if one wishes to understand the relationship between competencies and conceptions" (1994, p. 53). A defining characteristic of most, if not all, constructivist theories is some account of schema theory for modeling cognition.

COGNITIVE MODELING

Although connected by a common interest in number theory and a constructivist epistemological orientation, the studies in this monograph have been conducted within theoretical frameworks using qualitative methodologies particularly suited to their authors' own particular approach to investigating various cognitive and instructional dimensions of learning and teaching number theory. Thus each of these studies is self-contained and can be read independently of the others. For those who would prefer to read and reflect upon this monograph as a collective whole and would prefer to do so through a more global theoretical and methodological lens, we would like to suggest some means here for doing so.

Any attempt to provide a comprehensive framework for modeling the development of learners' understanding of number theory should, if at all possible, conform to the complexity that it aims to represent. Such a concern motivated Vergnaud to introduce and expound upon the notion of a *conceptual field* for modeling concept formation. "This complexity," according to Vergnaud, "comes mainly from the fact that mathematical concepts draw their meaning from a variety of situations and that each situation cannot usually be analyzed with the help of just one concept but rather requires several of them" (1990, p. 23). This was a conclusion we were led to in our studies of preservice teachers' understanding of divisibility and prime decomposition (Zazkis & Campbell, 1996a, 1996b). In those studies we drew upon Dubinsky's (1991) Action Process Object (APO) schema theory, or APOS, which he adapted from Piaget for the study of *advanced* mathematical thinking (see Brown et al., Chapter 3, this volume).

We found Dubinsky's theory particularly helpful for interpreting learners' understanding of various procedures and concepts pertaining to elementary number theory. We were also led to appreciate limitations in our ability to apply this framework. Our data "indicated complex interpenetrating layers of cognitive structure underlying the concepts of number theory that extend deep into the elementary operations and concepts of arithmetic" (1996a, p. 559): Arithmetic operations and concepts that are typically associated with additive and multiplicative structures, and that are intimately connected with informal knowledge and experience with everyday objects and activities (see, e.g., Mack, 1990). We concluded that "as the [APOS] framework is applied to more elementary concepts, the relationship between encapsulation [into objects] and thematization [into schemas] blurs into a host of 'cognitive substructures' " (Zazkis & Campbell, 1996a, pp. 561–562).

A more comprehensive approach to cognitive modeling of mathematical understanding ultimately requires some form of intuitive grounding in the more situated contexts of everyday lived experience (e.g., Lave, 1988). In developing his theory of conceptual fields, Vergnaud has noted that "Schemes [also] refer to situations; therefore it is essential to see the world of our experience as a set of situations, and not merely as a set of objects, qualities and relationships" (1996, pp. 224–225). Complementing, or possibly even incorporating, APOS theory with a theory oriented to the long-term development of learners' mathematical understanding that connects with the more situated and concrete kinds of experience underlying that understanding seems to be in order.

Briefly, Vergnaud's theory recognizes that concepts can rarely be meaningfully considered in isolation of each other, or the circumstances that have served to motivate their development. He has characterized conceptual fields as "large sets of situations whose analysis and treatment require several kinds of concepts, procedures, and symbolic representations that are connected with one another" (1990, p. 23). Vergnaud's notion of a conceptual field, as defined in terms of concept formation, is essentially comprised of the following: (1) a set of situations, the mastery of which serve to make concepts meaningful; (2) a set of operational invariants correlating with situational invariants that constitute concepts (e.g., APOS theory); and (3) a set of symbolic representations (natural language, icons, graphs, etc.) that refer to concepts, their properties, and the situations they refer to (cf. Vergnaud 1982, 1994, 1996).

An important reason for considering Vergnaud's theory of conceptual fields as a general thematic for viewing the studies in this volume is its generality. Vergnaud (1996) has drawn on a wide variety of cognitive theorists, such as Piaget and Vygotsky, as well as others, from Russell to Newell and Simon, in an attempt to provide a more comprehensive and overarching model of human experience and understanding. Thinking of number theory as a conceptual field is also appealing in that there seems to be a more natural concern in number theory, both historically and psychologically, with conceptualization than there is toward calculation and application. And Vergnaud has emphasized "the need to develop a theory that would make conceptualization the keystone of cognition" (1996, p. 224).

Perhaps most important, however, is that Vergnaud has already done much work in helping to resolve and define conceptual fields concerning additive structures (1982, 1996), multiplicative structures (1983, 1994, 1996), and elementary algebra (1996). The subsets of which, namely those pertaining and restricted specifically to whole numbers and integers, can readily be incorporated into a definition of number theory as a conceptual field.

It may prove helpful to the reader, therefore, to think about number theory as a conceptual field—though not so much as a common theoretical framework from which to interpret each of the studies in this monograph. That would require a degree of definition and standardization that has not been possible for the editors or the contributors of this monograph to achieve or adhere to at this point. Working *toward* a systematic definition of number theory as a conceptual field, however, is helpful and appropriate insofar as it can serve as a general thematic from which to view this monograph as a whole. Moreover, a thorough investigation and explication of number theory as a conceptual field can also be taken as a goal for a more comprehensive and systematic program of research in this area.

STUDIES TOWARD NUMBER THEORY AS A CONCEPTUAL FIELD

The area of study demarcated by number theory as a conceptual field (NTCF) may, as a first approximation, be considered as primarily concerned with the development of abstract procedural and conceptual understandings of the arithmetic and algebra of whole numbers and integers. The studies in this monograph can be taken as contributions toward a more fine-grained definition and explication of NTCF as an important and newly emerging area of research in mathematics education. Concepts studied in this monograph include division and divisibility as defined over whole and integer numbers, divisors, factors, and multiples, greatest common divisors and least common multiples, prime and composite numbers, and prime decomposition. Operations and procedures include prime factorization, finding of least common multiples, greatest common divisors, and determining relationships among interrelated concepts such as quotients and remainders. The main theorems include the division theorem (division algorithm), Wilson's theorem, various divisibility properties of integers ("divisibility rules"), and the fundamental theorem of arithmetic. Cognitive processes and concepts involved in reasoning with and proving these kinds of theorems over integers and whole numbers using proof strategies such as formulating conjectures and proof by induction are also included.

While various topics of number theory may be appropriate throughout the Pre-K–16+ curriculum, the studies in this monograph are focused primarily on cognition and instruction of undergraduates and prospective teachers, while covering a variety of different dimensions of learning and teaching number theory. Although many of these studies have been conducted in the United States, there are also international contributions from Canada, the United Kingdom, and Italy. Areas of interest in these studies range from questions of learning and cognition, to linguistics, to pedagogy, with much overlap in between. The topics of number theory used in these investigations range from questions that would not be out of place in the middle school curriculum to questions that constitute subject matter more appropriate to postsecondary levels.

To one extent or another, all of these studies are concerned with learners' approaches to problem solving, communicating, representing, connecting, and reasoning with introductory topics of elementary number theory, and these in turn have implications for the classroom. There has been no one predetermined way of classifying or sequencing these studies. The sequence that has been chosen flows reasonably well from studies concerned primarily with investigating learners' representation and understanding of number theory to studies concerned primarily with investigating learners' reasoning, particularly with respect to conjectures and proof. There has also been an attempt to balance a gradual shift from more introductory to more advanced topics in number theory, along with a transition from studies more concerned with cognitive content to studies more focused on instructional processes.

A set of situations or problem classes that give meaning to concepts is a key defining feature of Vergnaud's theory of conceptual fields. What makes the idea of NTCF so open for investigation is that little is known of what constitutes well-defined problem classes in this area. Thus far, however, it appears that learners tend to rely on invariant organizations of behavior (or schemas) acquired in other conceptual fields. This is most evident in Campbell's chapter, in which he investigates preservice teachers' understanding of division, where it appears that one of the main obstacles in students' understanding is their inability to readily distinguish between rational number division and whole number division with remainder. Campbell, in emphasizing that whole number division is defined by a basic theorem of number theory, provides some indication as to where the overlaps and boundaries between the multiplicative conceptual field (MCF) and NTCF may be sought and better resolved.

Brown, Thomas, and Tolias's contribution investigates the progression of prospective elementary teachers' thinking from action-oriented responses to problems concerning divisibility to more deliberate approaches that entail a more explicit understanding of mathematical operations and procedures pertaining to multiplicative structure of natural numbers. In so doing, they draw on a Piagetian progression from "success" (knowing by doing), to an intermediate reciprocal state between actions and concepts, to a more deliberate, reflective "understanding" (consciously guided action), which they attempt to integrate into APOS theory. Their investigation contributes to an ongoing effort to provide a more complete thematization of the concept of divisibility with respect to multiplicative structure.

Zazkis's and Ferrari's contributions shift from the cognitive to more linguistic and semantic concerns regarding divisibility and multiplicative structure. Zazkis builds on her previous investigations into a linguistic gap between the meanings of preservice teachers' informal natural language and formal mathematical language. She identifies and focuses on a variety of problems resulting from ambiguities associated with basic number theoretical terms associated with divisibility and multiplicative structure such as "divisible," "divides," "divisor," "factor," and "multiple." Zazkis analyzes different ways in which students may be motivated and constrained by their predispositions and preconceived meanings toward these terms.

Continuing along similar lines, Ferrari offers a study on the role of language in first-year undergraduate computer science students' understanding of some selected topics from number theory. Also drawing on a variety of previous studies in semiotics and mathematics education, he explores some differences between linguistic registers. In so doing, Ferrari introduces a construct he refers to as "semiotic control," which he considers in relation to APOS theory. This notion is of particular interest because it focuses on interpreting symbolic representations. Semiotic control speaks to the abilities of learners to interpret symbols in the sense of describing and making sense out of problem situations rather than as simple prompts for specific behaviors. Different levels of semiotic control are demonstrated in terms of the strategies that learners' use on problems that defeat the use of familiar procedures yet readily yield to strategies opened by more meaningful symbolic interpretations. The problems considered are drawn mostly from number theory, as Ferrari has found them particularly well suited for this kind of study.

All of the studies up to this point have been clinical studies focused primarily on modeling learners' cognitive understandings of subject matter content. Teppo's study is geared toward the integration of the development of conceptual content with processes of teaching in a classroom context. Again, the participants are preservice teachers and the classroom activity is based on a problem drawn from number theory that requires the students to investigate patterns in the number of divisors of natural (whole) numbers. In her reenactment of the classroom experience, Teppo illustrates how a variety of number theory concepts were implicated in a variety of mathematical processes and problem-solving activities. These classroom activities engaged learners in processes ranging from organizing information to making and testing conjectures. Teppo's contribution also provides a good indication of how teaching number theory can be facilitated by a pedagogy of "reflective discourse."

In a similar vein, Brown explores the problem-solving strategies of a variety of different learners in various contexts using an interesting problem from number theory that serves well to challenge some standard arithmetic preconceptions regarding the relationship between ordinality and representation. This problem, as with Teppo's and many other problems like them, help to illustrate ways in which prime decomposition and divisibility open up whole new levels of problems and opportunities for the development of students' structural understanding of mathematics. These kinds of problems evoke more basic levels of understanding of factors, multiples, and divisors, which often collide with the more familiar meanings of these terms with respect to the operations of arithmetic multiplication and division using rational numbers. Overall, the general result appears to be a greater appreciation and deeper understanding of arithmetic and the underlying properties and structures of whole numbers.

Collectively, the studies considered up to this point provide some indication of the potential that number theory has for helping learners recognize and redress limitations in their conceptual understanding of whole number arithmetic. The remaining three chapters of this monograph are more directly concerned with the potential of number theory for investigating learners' abilities to generalize and make conjectures to finding ways of justifying those conjectures with the development of inductive and deductive proof strategies.

Edwards and Zazkis's study explores preservice teachers' work on a problem situation, a complete solution to which requires incorporation of number theory concepts. Their study concerns ways that students respond to evidence that does not confirm a conjecture. Although there is a normative or mathematically correct response to such a situation, a study of preservice teachers suggested that not all students have internalized the norms that underlie this response. It also shows that students' difficulties are in part due to the fact that they often do not approach problems in number theory in a number theoretical frame of mind. A similar observation can be made about Brown's study—students' obstacles in tackling the problem are, at least in part, due to the fact that the problem they are presented with is not approached as situated within NTCF.

Rowland explores NTCF in terms of its stimulating theorems and their proofs. His focus is on "generic examples" that provide the main structure without formalism, which may, in his pedagogical perspective, serve as a springboard to a rigorous proof. And, finally, Harel's study focuses on one important technique applied in number theory proofs—mathematical induction. He explores mathematical induction as a proof scheme, elaborates on an instructional treatment implemented in a number theory course, and describes how a fundamentally different pedagogical approach helped students in developing their proof schemes.

CONCLUDING REMARKS

Philosophers and mathematicians since the time of Pythagoras have believed that understanding number provides the key to understanding the world. Indeed, this fundamental notion remains essentially unchanged to this day in the guise of mathematical modeling. The main difference is that we now have a much more general concept of number than did the ancients. Today it is common practice to consider whole numbers, for all practical intents and purposes, as a subset of rational numbers (see, e.g., Freudenthal, 1983). This practice carries with it the implication that any particular whole number is also a rational number and therefore can, both in principle and in practice, be divided. To subsume one to the other, however, is to lose sight of one of the most profound conceptual shifts in the history of mathematics: the shift from indivisible to divisible units (Klein, 1968/1992) and all of the many important formal mathematical implications that follow from making such a distinction.

A key question to resolve, therefore, is the extent to which learners may benefit from having an abstract number theoretical understanding of arithmetic and algebra using integral quantities prior to, or at least in clear contrast with, learning arithmetic and algebra with nonintegral quantities. Studies in this monograph indicate that conflating the two often presents learners with a wide variety of procedural, conceptual, and linguistic difficulties. The extent to which developing clear and meaningful conceptual understandings of properties and structures of integral quantities may contribute to improved algebraic understandings of notions such as variables, functions, and of problem solving and reasoning with nonintegral quantities, remains to be seen. Overall, studies in this monograph have taken important steps toward eventually addressing these kinds of questions, but much work remains to be done.

The extent to which learning and teaching number theory may alleviate or amplify some of the many well-known problems involved in helping teachers and learners to develop a more profound understanding of fundamental mathematics also remains to be seen. The potential that learning and teaching number theory has for improving mathematics education can hardly begin to be charted or fathomed without a comprehensive and systemic effort on the part of the community of mathematics educators and researchers. Such an effort has yet to be made, and such an effort is not to be found here. The task is too big. Our aspirations for this monograph are more humble. Research to date in this area has been relatively sparse and disconnected. If the guiding theme of understanding number theory as a conceptual field, and the following studies working toward that end reported herein, help to convince researchers and educators alike that more work in this area is warranted, then we have achieved our purpose.

REFERENCES

Dörfler, W. (1995). Mathematical objects, representations, and imagery. In R. Sutherland & J. Mason (Eds.), *Exploiting mental imagery with computers in mathematics education* (Vol. 138, pp. 231–247). Berlin: Springer-Verlag

Dubinsky, E. (1991). Reflective abstraction in advanced mathematical thinking. In D. Tall (Ed.), *Advanced mathematical thinking* (pp. 95–123). Boston: Kluwer Academic.

Durkin, K., & Shire, B. (1991). Lexical ambiguity in mathematical contexts. In K. Durkin & B. Shire (Eds.), *Language in mathematical education: Research and practice* (pp. 71–84). Milton Keynes, UK: Open University Press.

Freudenthal, H. (1983). *Didactical phenomenology of mathematical structures*. Boston: D. Reidel.

Glasersfeld, E. v. (1982). An interpretation of Piaget's constructivism. *Review Internationale de Philosophy, 142 (3)*, 612–635.

Glasersfeld, E. v. (1995). *Radical constructivism: A way of knowing and learning*. London: Falmer Press.

Greer, B. (1987). Nonconservation of multiplication and division involving decimals. *Journal for Research in Mathematics Education, 18*(1), 37–45.

Hardy, G. H. (1967). *A mathematician's apology*. London: Cambridge University Press. (Original work published 1940)

Healy, L., & Hoyles, C. (2000). A study of proof conceptions in algebra. *Journal for Research in Mathematics Education, 31*(4), 396–419.

Kant, I. (1965). *Critique of pure reason* (N. K. Smith, Trans.). (Unabridged ed.). New York: St Martin's Press. (Original work published 1787)

Kieran, C. (1992). The learning and teaching of school algebra. In D. A. Grouws (Ed.), *Handbook of research on mathematics teaching and learning: A project of the National Council of Teachers of Mathematics* (pp. 390–419). New York: Macmillan.

Klein, J. (1992). *Greek mathematical thought and the origins of algebra*. New York: Dover. (Original work published 1968)

Kornblith, H. (Ed.). (1994). *Naturalizing epistemology*. Cambridge, MA: MIT Press.

Lave, J. (1988). *Cognition in practice: Mind, mathematics and culture in everyday life*. Cambridge, UK : Cambridge University Press.

Ma, L. (1999). *Knowing and teaching elementary mathematics: Teachers' understanding of fundamental mathematics in China and the United States*. Mahwah, NJ: Erlbaum.

Mack, N. (1990). Learning fractions with understanding: Building on informal knowledge. *Journal for Research in Mathematics Education, 21*(1), 16–32.

Mack, N. (1995). Confounding whole-number and fraction concepts when building on informal knowledge. *Journal for Research in Mathematics Education, 26*, 422–441.

Mathematical Sciences Education Board. (1989). *Everybody counts: A report to the nation on the future of mathematics education*. Washington, DC: National Academy Press.

National Council of Teachers of Mathematics. (1989). *Curriculum and evaluation standards for school mathematics*. Reston, VA: Author.

National Council of Teachers of Mathematics. (1998, October). *Principles and standards for school mathematics: Discussion draft.* Reston, VA: Author.

National Council of Teachers of Mathematics. (2000). *Principles and standards for school mathematics.* Reston, VA: Author.

Piaget, J. (1972). *The principles of genetic epistemology.* London: Routledge.

Sfard, A. (1991). On the dual nature of mathematical conceptions: Reflections on processes and objects as different sides of the same coin. *Educational Studies in Mathematics, 22*(1), 1–36.

Silver, E. A. (1992). Referential mappings and the solution of division story problems involving remainders. *Focus on Learning Problems in Mathematics, 14*(3), 29–39.

Vergnaud, G. (1982). Cognitive and developmental psychology and research in mathematics education: Some theoretical and methodological issues. *For the Learning of Mathematics, 3*(2), 31–41.

Vergnaud, G. (1983). Multiplicative structures. In R. Lesh & M. Landau (Eds.), *Acquisition of mathematics concepts and processes* (pp. 127–173). New York: Academic Press.

Vergnaud, G. (1990). Epistemology and psychology in mathematics education. In P. Nesher & J. Kilpatrick (Eds.), *Mathematics and cognition: A research synthesis by the International Group for the Psychology of Mathematics Education* (pp. 14–30). Cambridge, UK: Cambridge University Press.

Vergnaud, G. (1994). Multiplicative conceptual field: What and why? In G. Harel & J. Confrey (Eds.), *The development of multiplicative reasoning in the learning of mathematics* (pp. 41–59). Albany: State University of New York Press.

Vergnaud, G. (1996). The theory of conceptual fields. In L. P. Steffe, P. Nesher, P. Cobb, G. A. Goldin, & B. Greer (Eds.), *Theories of mathematical learning* (pp. 219–239). Mahwah, NJ: Erlbaum.

Zazkis, R., & Campbell, S. R. (1996a). Divisibility and multiplicative structure of natural numbers: Preservice teachers' understanding. *Journal for Research in Mathematics Education, 27*(5), 540–563.

Zazkis, R., & Campbell, S. R. (1996b). Prime decomposition: Understanding uniqueness. *Journal of Mathematical Behavior, 15*, 207–218.

2

Coming to Terms with Division: Preservice Teachers' Understanding

Stephen R. Campbell

A number of studies have investigated preservice teachers' understandings involving abstract problem tasks in areas such as division of, and by, zero (Ball, 1990; Wheeler & Feghali, 1983), long division, division with remainder using calculators (Simon, 1993), and divisibility (Zazkis & Campbell, 1996a). Studies such as these are in accord with the National Council of Teachers of Mathematics' (NCTM, 1989, 2000) emphasis upon the importance of investigating the connectedness of learners' content knowledge. This study[1] contributes to this line of research, and is another component of a more general investigation into preservice teachers' understanding of elementary number theory (Campbell, 1996; Campbell & Zazkis, 1994; Zazkis & Campbell, 1996a, 1996b). The main focus of this study is on preservice teachers' understandings of concepts, procedures, and terms pertaining to "the division algorithm," a fundamental theorem of number theory that defines whole (and integer) number division. Typically referred to by mathematicians as "the division algorithm," this theorem will be referred to here as "the division theorem" to ensure it is not confused with the algorithm for long division (see "Whole Number Division" in the Appendix).

Thus more specifically, this study investigates various aspects of preservice teachers' content knowledge related to the division theorem using abstract problem tasks involving prime decomposition and calculators. These problem tasks were designed to be readily resolved with an adequate understanding of the basic concepts involved. However, the preservice teachers participating in this study encountered a wide variety of difficulties when presented with these tasks.

There was a general propensity on the part of these participants to misapply familiar procedures in novel or unfamiliar circumstances. Other difficulties involved gaps in understanding and uncertainty regarding the meanings of numerical terms, expressions, objects, and operations of division, and a variety of connections between them.

BACKGROUND AND MOTIVATION

Division is a multifaceted notion that is subject to many closely related, but quite distinct, interpretations in mathematics. Although intuitively evident to children and adults learners alike as partitioning a concrete quantity into a given whole number of portions (Fischbein, Deri, Nello, & Marino, 1985), there are formal conditions and restrictions on division pertaining to the domains over which it is defined (see, e.g., Freudenthal, 1983). There are, of course, underlying similarities connecting various aspects of division such as divisibility, whole number division, and rational number division (see Appendix). This study, however, suggests greater discernment is warranted with respect to some of the fundamental procedural and conceptual differences between them. This study is primarily concerned with preservice teachers' understanding of these differences and some of the difficulties they encountered in coming to terms with them.

A previous report drawing upon data selected from the same interviews as this study revealed difficulties in preservice teachers' understanding of divisibility and the multiplicative structure of natural numbers (Zazkis & Campbell, 1996a). Although beyond the scope of that report, evidence was presented therein suggesting the participants' understanding of division was implicated in some of those difficulties. Responses to the following question were particularly telling: "Consider the numbers 12,358 and 12,368. Is there a number between these two numbers that is divisible by 7?" Results obtained ranged from procedural approaches of dividing, using either a calculator or long division, until the desired number was found, to more conceptual approaches that gave rise to appropriate answers with little or no computation whatsoever. Between these extremes were variations suggesting difficulties in understanding whole number division in relation to, and as distinguished from, divisibility and rational number division. In using either a calculator or long division to address this question, and other related questions from these interviews (below), participants confronted a variety of obstacles and were led to a variety of inferences involving various aspects of division.

For example, one can use a calculator to evaluate $12{,}359 \div 7$ to obtain, depending on the precision of the display, say 1265.5714286 (rational number division), or one can use long division to evaluate the same expression to obtain a quotient of 1265 with a remainder of 4 (whole number division). Now, there was no indication that any of the participants in this study were unaware that for one number, A, to be divisible by another, D, entails D must divide A "evenly." This "fact" means, depending on the approach taken, that either the fractional component of the rational quotient or the whole number remainder must be zero. With either of these basic connections between divisibility and division, one can simply continue evaluating numbers between 12,358 and 12,368 until one either exhausts the possibilities or identifies a number divisible by 7.

There are, however, other connections between divisibility, whole number division, and rational number division that can be brought to bear on the solution of this problem. For instance, after using long division to obtain a remainder of 4, one of the participants in this study, Jane, was able to draw on her understanding of whole number division by "counting on" from 4 to 7 to infer that 12,359 + 3 will be divisible by 7 (Zazkis & Campbell, 1996a, p. 548). Another participant, Pam, in foregoing any calculation whatsoever in addressing this problem, provided evidence of a more sophisticated view of numbers divisible by 7 based, at least implicitly, on an understanding of the modular distribution of the multiples of 7 (p. 554). Below, it will be seen that Pam's understanding of the connection between the concepts of divisibility and multiple was based, at least implicitly (i.e., as a "theorem in action," Vergnaud, 1996), on an understanding of the division theorem.

Moreover, identifying divisibility with "dividing evenly," that is, either with a zero remainder or fractional component, is but a special case of a more general relationship between the remainder of whole number division and the fractional component of rational number division. As we shall see, the connection between the two was not readily evident to the participants in this study without some knowledge, again at least implicitly, of the division theorem.

METHOD

One of the most important distinctions in arithmetic division is that between whole number and rational number division. This distinction entails a major shift in the meaning of division along with other associated concepts, procedures, and terms involved. The main avenues taken here for analyzing the connections between these shifts in meaning were problems using novel forms of expression for the dividend, whole number division using a calculator, and partitive and quotitive dispositions toward division. These differences, and the various interpretations and associated symbolism used for representing them, can be found in the Appendix.[2]

A wide variety of responses to the problem tasks were obtained and various considerations are raised in order to account for them. Theoretical considerations have been introduced when appropriate for modeling procedural errors, and for identifying potential sources of conceptual obstacles and linguistic ambiguities. Specific examples and general patterns pertaining to such errors, obstacles, and ambiguities are presented and discussed. The methodological stance of this study is oriented toward qualitative descriptions and interpretations of phenomena that may commonly, but not necessarily, be exhibited in preservice teachers' understandings of division.

Participants and Protocol

Individual clinical interviews were conducted over a two-week period in mid-semester with twenty-one volunteer preservice elementary teachers from a group of students enrolled in a professional development course called "Foundations of Mathematics for Teachers." The mathematical background and experience of the participants varied considerably. The course curriculum treated by the time of the interviews were conducted had covered basic topics from elementary number theory such as divisibility, factorization, prime decomposition, and so on. Divisibility rules for the numbers 2, 3, 5, and 9 were also part of their curriculum, although an explicit treatment of the division theorem was not.

The interview protocol allowed for the flexibility to probe and clarify participant understandings of the procedures and concepts involved in the problem tasks presented. Occasionally, as warranted by the context and constraints of the interview, not all questions were asked of all participants and some questions were not asked in the same order for reasons that are indicated in the few cases where this occurred. Calculators were available for the use of the participants. Question sets from these interviews, pertaining to divisibility, divisibility rules, factorization, and prime decomposition, have been analyzed, interpreted, and reported elsewhere (Zazkis & Campbell, 1996a, 1996b). Pseudonyms for participants used in these previous reports are the same as those used herein.

Interview Questions

The question sets were primarily designed to present familiar concepts in novel contexts. Overall, the instrument served to guide the investigation of content knowledge affecting each participant's ability to assimilate and/or accommodate these problems. The question sets from the interviews pertaining to this study are:

Question set 1: If you divide 21 by 2, what would the quotient be? What would be the remainder?

Question set 2: Consider $M = 3^3 \times 5^2 \times 7$. If you divide M by 15, what would the remainder be? What would be the quotient?

Question set 3: Suppose you're asked to perform division with remainder on 10561 divided by 24. Will your calculator help you? How?

Question set 4: Consider the number $6 \times 147 + 1$, which we will refer to as A. (a) If you divide A by 6, what would be the remainder? What would be the quotient? (b) If you divide A by 2, what would be the remainder? What would be the quotient? (c) Consider the number $6 \times 147 + 2$. When it is divided by 2, what is the remainder? What is the quotient?

Question set 5: What does it mean to you to say that division is an inverse of multiplication? Is there any way in which you see division with remainder as being an inverse to multiplication?

The first question set was designed solely with the intent of ensuring participants could properly relate linguistic terms with the intended whole number values associated with the quotient and remainder. Although serving its initial purpose, an unanticipated result has emerged regarding the potential relevance of partitive and quotitive dispositions toward division with respect to the domains of whole and non-negative rational numbers. Although no systematic attempt was made to evoke participant dispositions in this regard, enough data were obtained to suggest their relevance. Question sets two through four were designed in the spirit of the NCTM (1989, 2000) curriculum standards, which emphasize the importance of evaluating integration and connectedness of concepts and procedures in novel and unfamiliar contexts. Note that in each case, quotients and remainders can be obtained by novel means from alternative expressions of a whole number dividend. Question sets three and five were designed to explore ways in which the participants were able to make connections and reconcile differences between whole number and rational number division. Question set four was designed to explore participants' tacit understandings of the division theorem.

Data Reduction

The interviews were recorded and transcribed from audiotape. Each of the question sets for each interview were analyzed, encoded, and, where possible, cross-referenced and compared with respect to the wide variety of responses expressed by the participants. There were some responses that, either due to lack of clarification regarding the language used or the approach taken, are not considered here. Those data for which some plausible observations, interpretations, or hypotheses could be drawn are considered below. Although frequency of occurrence for some of these phenomena are noted, there are no implications intended regarding their statistical significance.

RESULTS AND INTERPRETATION

Partitive and Quotitive Dispositions

Of the 21 participants considered in this study, 16 were asked to provide the quotient and the remainder of 21 divided by 2. This question was intended to clarify whether participants understood the reference of the terms quotient and remainder. Most participants (12 of the 16) readily provided answers for the quotient as 10 and 1 for the remainder. Responses obtained from Jim and Anabelle respectively illustrate two basic facts about division with whole numbers that can readily be taken for granted: first, that whole number division requires that the remainder and quotient are whole numbers; second, that the remainder be smaller than the divisor. More importantly, however, these cases serve to indicate the potential relevance of partitive and quotitive dispositions in understanding arithmetic division in abstract contexts.

Partitive Dispositions and Rational Numbers

Jim's initial response for the remainder of 21 divided by 2 was .5. An analysis of the transcribed data with the worksheet he used through the course of the interview confirmed that he performed long division to obtain a rational quotient of 10.5 and identified .5, the fractional component of the rational quotient, as the remainder. When the interviewer asked Jim how he would conceive the remainder as a whole number he responded as follows:

Jim: We're dealing with 2 parts, so we'll add them together and get 1, remainder 1.

Interviewer: Okay. When you say that we're dealing with 2 parts, would you explain that a bit?

Jim: We're dividing the 20 into 2 parts. [. . .] So you're taking the 21 and, [I] don't know, it has a picture, you're taking it and splitting it in half and finding out what fits on either side.

Interviewer: Oh, so you've got 10 1/2 in here . . .

Jim: So you've got 10 parts here and 10 parts here, but you've still got that one part so half of it is over here, and half of it is over there. [. . .] So you've got 2 half parts, so you add them together, and you get 1.0. [. . .] And you get remainder 1. Look at it either way.

This excerpt indicates that Jim considered both division and the arithmetic units involved partitively. When asked for an account of the remainder in terms of whole numbers, Jim is evidently expressing "what fits on either side" in whole units. He views the quotient in terms of the magnitude of the whole number "parts." However, in order to make sense of the one unit left that doesn't "fit," Jim divides again. But to do so the remaining unit must be conceived as a divisible (rational number) unit rather than as an indivisible (whole number) unit. The "either way" to look at the remainder that Jim refers to can be interpreted with respect to a whole number remainder (had he *not* divided the "part" that didn't fit in his initial partition) or as a "part," or unit, that can be divided once again and added into the initial partition. Either way, Jim's disposition here toward whole number and rational number division is a partitive one.

Despite the fact that the problem situations in this study were presented in the context of elementary number theory in the domain of whole numbers, at least two-thirds of the 21 participants in this study, at one point or another, expressed remainders or quotients as rational numbers. Most of these cases led to considerable confusion, often impeding or preventing participants from achieving problem task solutions. As we shall see, many of these cases appear to be connected with partitive dispositions toward division.

Quotitive Dispositions and Whole Numbers

With respect to the constraint that the remainder be smaller than the divisor, Anabelle initially stated the remainder was 3 with a quotient of 9, noting "2×9 is 18, plus 3 is 21." When the interviewer intervened to point out that the remainder was meant to be understood as a quantity less than the divisor, she responded:

Anabelle: Right, I'm wrong. 10, with remainder 1 [*laugh*]. I did that same mistake on the exam. Common error for me.

Interviewer: Why do you think you have a tendency to look at it that way?

Anabelle: Yeah, I went with the multiple closest, but the multiple closest would have been 10. I thought 9 was the closest to it, because of 18, I never even thought of 20.

Anabelle's quick reaction to the intervention and subsequent explanation indicates that her initial response pertained more to her familiarity with, or recall of, specific number facts rather than a misconception regarding the constraint upon the remainder. Fortunately, this event opened a window to her disposition toward whole number division. In contrast to Jim, Anabelle did not perform long division in order to find the quotient and remainder. In looking for the "closest" multiple, she may have at least an implicit awareness of quotitive division as the number of times the divisor "goes into" the dividend. At least ten of the twenty-one participants in this study, on different occasions and problem contexts, seemed similarly disposed toward a quotitive understanding of division. Again, as we shall see, most of these cases led readily to problem task solutions.

Contextual Ambiguities and Procedural Attachments

Nineteen of the twenty-one participants in this study were asked what they thought the remainder would be if $M = 3^3 \times 5^2 \times 7$ was divided by 15. Of these nineteen participants, eight eventually assimilated the problem by calculating the dividend in order to determine the remainder, while only four, appealing to various divisibility criteria, were able to respond definitively that the remainder would be zero. Insofar as these responses pertain to understandings of divisibility and prime decomposition, they have been considered elsewhere (Zazkis & Campbell, 1996a, 1996b).

Of interest here is that at least five of the nineteen participants asked this question were unsuccessful in their attempts to either assimilate or accommodate this novel experience. These participants all confused the remainder with the quotient in terms of "what you are left with." Three of these five were subsequently asked about the quotient. Two participants identified the quotient as the divisor while one identified the quotient as 1. The following excerpts from the interviews with Riva and Lanny illustrate these cases with respect to the remainder:

Interviewer: What would you expect the remainder to be, if you were to divide M by 15?

Riva: Um, [*pause*] you're probably left with a 3^2, a 5, and a 7.

Interviewer: And if you wish to divide M by 15, what would you expect the remainder to be?

Lanny: [*Pause*] Divide that by 15, um, well it's going to divide evenly because the, it is, like it's a factor of, like a multiple, I don't know if it's a factor or not. Um, because it has 3 and 5, so 3×5 is 15, um, but then there's the remainder be, um, because you have 3^3 and 5^2 so, like you're still going to have 3^2, a 5, and a 7, so you're only using like one 3 and one 5, so you like well for sure you'll have, well that'll be your remainder, I guess . . . .

Evidently, these participants had difficulty discerning differences or making connections between divisibility and whole number division with remainder equal to zero. It is likely that such thematization is required in order to accommodate this shift in application. What positive accounts can there be for these responses? It could be that some participants were guided by an expectation that there must be a remainder, simply because the interviewer had posed the question. However, such an explanation seems inadequate in these cases. Without some conceptual understanding or structural insight for making thematic connections and distinctions in novel contexts, individuals often attempt to apply or adapt known procedures from more familiar contexts (Matz, 1982).

Both Riva and Lanny were able to successfully divide by factoring 15 from M. The difficulty that confronted them then was interpreting their results in this abstract context. After obtaining the above responses, the interviewer, in both cases, inquired what they would consider the remainder and quotient to be upon dividing 21 by 2. Both Lanny and Riva responded without hesitation that the remainder would be 1 and the quotient would be 10. It was not simply a matter of not understanding the meanings of the terms involved in a more familiar context. The interviewer, then, in both cases, returned to ask once again about the remainder, and subsequently what the quotient would be, in dividing M by 15.

Riva's approach to the problem seemed unaffected by the intervention. She indicated that the quotient "would probably be 1" because that was the number of times she had factored the divisor out of the prime decomposition of M in order to obtain the "remainder" in terms of "what's left." Could Riva have been treating the prime decomposition analogously to the decimal representation of the dividend and imagining the result of performing long division? This interpretation could explain how she obtained a quotient of 1 and was "left with a 3^2, a 5, and a 7."

Lanny's response was more involved as she struggled to make some sense of this novel situation in terms of a context she was more familiar with:

Interviewer: Okay, could you use that same method [used for 21 divided by 2] then with M?

Lanny: M? Yep. Um, [*pause*] it goes, it's $3^2 \times 5^2 \times 7$, right, and then you said it's 15, divides into that, so I took 15 and I [have] 3×5, so we have our number here, 3×5 so um, okay, we'll look at this number as, I'll write it out ??, 3 . . . no this is cubed, $3 \times 3 \times 3 \times 5 \times 5 \times 7$ and so it's one 3 and one 5, so you take out ?? this 3 and this 5 and so then you get, your remainder is $3 \times 3 \times 5 \times 7$, it's not exactly the same because here [21 divided by 2], like your taking, isolating the 10s, divide, it's sort of the same because here you're isolating what 15 represents, 3 and the 5.

Interviewer: And what would then be the quotient for that problem?

Lanny: Hmm, for this M one? [*Pause*] 3×5, I guess, but then, yeah, because the remainder is $3^2 \times 5 \times 7$, but the quotient, I don't know. The way it's set up like this, I don't know. That's a good question [laugh]. Um, because I, you know, following this suit, the way this is set up, what I took out was my quotient, because I took out a 10, so I'd prob . . . I would divide it, so if you asked me on an exam, I would say I guess the quotient was 3×5 because that's what I was taking out from this. But then you're asking me to divide this whole thing by 15, and then here I'm saying the quotient is 15, I don't know. I don't understand, I thought I understood the meaning of quotient, [. . .] when it's set up like that I don't know how to take that meaning and apply it to that situation.

Although these data are not readily interpretable, some observations seem warranted. Throughout her interview, Lanny consistently demonstrated procedurally oriented responses to the questions asked of her. She also had a propensity to view multiplicative problems additively. For her, prime decomposition was related to "breaking down" a number into a factor tree and she understood divisibility as a matter of factoring—in terms of "isolating" and "taking away." Apparently, Lanny was "subtracting" and an imprecise understanding of the remainder may have led her to conclude that it referred to what "you get" after factoring. This interpretation is consistent with her subsequent attempt to make a connection between the quotient and subtrahend in terms of what she was "taking out" (a "share"? was she thinking of division partitively?). However, Lanny notes an apparent conflict of that assignment with the divisor, which seems to undermine her confidence that she can successfully extend her understanding of quotient to this situation.

These cases, in keeping with Matz's (1982) emphasis on processes where familiar activities and rules are mistakenly applied or extended to novel tasks and situations, appear to involve inappropriate extensions of meaning as well. Linda provided another good illustration of this tendency. In response to the question asking for the remainder of $6 \times 147 + 1$ when divided by 6, she said: "147 + 1, um, I came to that just because, if you're multiplying something by 6 and then dividing it by 6, it cancels out the number 6, therefore, the remainder that you have is 147 + 1." This serves as an instantiation of a common algebraic error: $(AX + B)/X = A + B$. This particular error is combined with a common, though imprecise, understanding of multiplication and division as inverse operations, namely, in terms of "cancellation."

In yet another case, Dana avoided Linda's procedural error in this situation, dividing both addends by 6, obtaining $(6 \times 147/6) + 1/6$, but was stymied in that she could not relate the meaning of the fractional addend, 1/6, to a whole number remainder. Noting 6 as a common denominator, she calculated 883/6 and then performed long division. More than half our participants experienced analogous difficulties assimilating or accommodating the meanings of quotient and remainder in these disequilibrating abstract contexts of question sets two, three and four—more instances of which will be presented below.

Whole Number Division Using Calculators

One way of investigating how effectively differences between whole and rational number division are understood is through tasks that require learners to explore the relationships between the abstract concepts involved. A case in point to this end is whole number division using calculators.

Seventeen of the participants in this study were asked how their calculators could help them to determine the remainder in dividing 10,561 by 24. Seven participants, using the fractional approach, eventually concluded that a calculator would be of no help to them whatsoever. Four of these seven participants were aware that in most cases division with a calculator would result in a value to the right of the decimal point that would be "completely different" from a nonzero whole number remainder and claimed to see no relationship between them. Some participants indicated that there was "a way to do it" but they just couldn't remember or didn't know what it was.

Fractional Approaches

Out of the ten participants that fixed their attention upon the fractional component of the rational quotient, *not one* was able to use the calculator to identify the exact whole number remainder. The difficulties and limitations of the fractional approach gave way in almost all cases to these participants either assimilating the problem by resorting to long division to determine the remainder or giving up altogether.

Some participants noted that the fractional component of the calculated rational quotient was not the desired whole number remainder. For instance, upon prompting, Lena tried to explain the problem by considering the case of 10 divided by 3:

Interviewer: Um, do you think there is something that you could do to find the remainder from that answer that you get from the calculator?

Lena: Well, [*pause*] let me see something. I'm thinking how many times, I'm just trying to think of an example, so 3 goes into 10, 3 times, the remainder 1, and on a calculator you're not going to have like 3.1 on a calculator.

Bob's response was virtually identical, noting that "you won't actually see the remainder on the calculator" because the latter "breaks it down to 1/10ths and 1/100ths." In a similar vein, Jennifer, upon comparing the remainder obtained after dividing by long division with the fractional component obtained from the same division with the calculator, exclaimed: "I can't for the life of me figure out why the calculator lied to me." Another participant, Tara, confessed to having no understanding of the relation between fractions and decimals.

On the other hand, Stanley was one of only two participants that came "close" to successfully determining the whole number remainder using the fractional method. Note that Stanley makes sense of the fractional component by first recalling the fractional approach. Then, realizing that the remainder must have been divided by the divisor as well, he inverts the procedure to obtain a value "very close" to the whole number remainder:

Interviewer: Suppose you wish to perform division with remainder, and here's the division: um, 10,561 divided by 24, now if, if we wanted to get the remainder from that, um, would your calculator help you?

Stanley: Sure, ?? [*pause*] okay, and it leaves me a decimal?

Interviewer: Um hmm.

Stanley: Uh, .04166 and I times that by 24, [*pause*] because, right because the remainder would be R, divided by 24 uh would equal .04166, yeah so I times it by 24 and that gives me the remainder.

Interviewer: And the remainder in this case is?

Stanley: Uh the remainder in this case is very close to 1 . . .

Interviewer: Um hm.

Stanley: So it goes .999 whatever.

Interviewer: Okay. Um, just an aside here, but let's say we took .999 and took 9s indefinitely, to infinity, do you think that number would necessarily be exactly the same as 1?

Stanley: Uh, well as the 9s go to infinity uh it becomes closer and closer to one, but it'll never actually reach one.

These excerpts serve to illustrate the broad spectrum of pitfalls and subtleties involved in determining a whole number remainder from a fractional approach that conflates numerical domains. Most of the other participants adopting a fractional approach were confronted, in various degrees, with the "friendliness" of the decimal expansion involved. Anita's experience offers a good case in point:

Interviewer: So if you wanted to know the remainder, is there anything that you could think of that you could do to change that decimal to a whole number remainder?

Anita: Whatever tha, if the decimal isn't um infinitive [*sic*], like never ending, then, if it's just like a .25 or whatever, you can put, or whatever it is, you can put the number over 10, 100, 1,000, 10,000, whatever, it's the corresponding um base, like 10 basis, 10 ??

Interviewer: Hmm.

Anita: Like if it was .25, if it was a number .25 then you just put .25, you put 25 over and that's 1 quarter, so remainder 1, but that would be if it was divided by 4, I guess.

The calculator was of use to many participants only to the extent that they were able to recognize, or make some sense out of, the decimal expansion of the fractional component of the quotient obtained from the division. Following along Anita's line of thought, the "friendliness," or familiarity of learners with particular patterns, of the decimal expansion is related to the divisibility relations that obtain between divisor and remainder with powers of 10, the representational base. Consequently, cases of division involving divisors such as 2, 4, and 5, etc., would be considered more friendly than divisions involving divisors such as 7, 11, and 12, etc., as the former are divisors of lower powers of 10 and the latter are not. A possible exception regarding this criterion would be the case of division by 3 due to the peculiar patterns and numbers involved in the expansions. One way of imposing friendliness upon unfriendly divisors would be to truncate the decimal expansion of the resultant quotient to a smaller power of 10. This variation of the fractional approach, however, confronts the learner with more serious round-off effects than that which Stanley contended with above.

Integral Approaches

In contrast, the experience of seven of the eight participants who used an integral method suggests that it is, conceptually and procedurally, a much more accessible approach to whole number division using a calculator. Jane was a participant who struggled valiantly with a variety of fractional approaches. Although she eventually did come fairly close to a procedural solution of the problem, Jane's truncation of the decimal representation of the fractional component of the rational quotient at an early point in her calculations seriously complicated the problem for her. It compromised her ability to obtain the integral solution required. However, after Jane had derived the division theorem "a = cb + d" from "a ÷ b = cRd" in response to question set five, she was invited to reconsider the problem:

Interviewer: When you have this written in front of your eyes, it doesn't give you any clue of what you may be doing here, once again, our example, 10,561 divided by 24 . . . ?

Jane: I need a product of b, so this 10,561 would equal 24 multiplied by 440 plus 4,167 100,000ths.

Interviewer: Plus whatever, plus the remainder that we are looking for . . .

Jane: Um hm.

Interviewer: Do you want to try to do this?

Jane: Sure. [*pause*] So 24 multiplied by 440 + [*pause*] I'm left with 1 . . .

Interviewer: So you did this multiplication and you got 10,560.

Jane: Um hmm. Which would mean that my remainder is one.

Interviewer: How? From here you ?? that your remainder is 1 . . .

Jane: Because 10,561 subtract 10,560 is equal to 1.

Note that Jane continued to identify the fractional component of the rational quotient with the whole number remainder. With further prompting, Jane obtained the whole number remainder using an integral approach. With the exception of Andy (considered in detail below), the other participants taking an integral approach had no evident difficulties at all. The following excerpts are quite typical:

Pam: 10,561 divided by 24, that's 440 with a decimal, so it's [*pause*] ??, there's a decimal there, so then I take that 440 and multiply that by 24 . . . times 24, which is equal to 1, 5, . . . no, 10,560 and you take the original number, minus that, which is 1.

Armin: 10,561 divided by 24 equals 440.04167 . . . 24 goes in here 440 × 24, so you go 10,560, with remainder 1.

Karen: Um, well the way, actually when I use a calculator to do um, to try things like this with larger numbers, um, you have to use it in a different way for it to help you, like you have to find out how many times; what I do with the calculator is, I find out how many times 24 goes evenly into 10,561 and then I multiply that number by 24, I see what the number, um, that comes up with it, it'll be slightly less . . . either it will be even, it'll be the same as this, or slightly less, and you subtract um, you

> subtract the number [product of quotient and divisor] it [the calculator] comes up with from this [the dividend] to get the remainder.

Apparently, an integral approach to whole number division with calculators seems virtually effortless when combined, implicitly or explicitly, with a quotitive disposition toward whole number division. Although the integral approach is structurally related to quotitive division and the division theorem, it does not follow that a procedurally oriented integral approach is sufficient to make the connections. This is illustrated by Andy's experience.

Andy encountered significant difficulties using the integral approach. Things began well as she used the calculator to divide 10,561 by 24 to obtain the rational quotient. Andy then multiplied 24 × 440 and it seemed as though her approach was on track. However, she then divided 10,561 by 440 to obtain 24.00227 . . . and referred to this new quotient as "a remainder of 24 [with a] decimal point afterward, so it didn't go in evenly." Analysis of the interview failed to clarify what may have motivated Andy to divide the dividend by the integral part of the rational quotient. It is apparent, though, that her focus at this point was upon divisibility rather than division with remainder.

In an attempt to encourage Andy to reconsider the problem, the interviewer provided a simple example with the hope of clarifying the task at hand: "for example, if I wish to divide 11 by 2, I can write my answer as 5.5, okay I can write my answer as 5, remainder 1, and I am interested in this kind of representation, the second one." Andy tried the problem again, dividing 10,561 by 24, stating "Um, it's 440, remainder something." She noted that [the whole number quotient] 440 couldn't be the remainder, and then claimed it was 441. Responding to a prompt for further clarification, Andy multiplied 441 by 24 and exclaimed: "Oh that doesn't work." She then concluded: "Okay, it's not 24 times 440, it's 110, 10,560, remainder 1."

Understanding the Division Theorem

Andy's reasoning, despite being procedurally oriented, seems incomprehensible. However, it is evident that at various points along the way she: (1) added the quotient to the remainder, (2) multiplied the divisor with the sum of the quotient and remainder, and (3) identified the product of the divisor and quotient with the quotient. It was not evident at any point in the interview that Andy was ever thinking of 440 as the quotient. Be that as it may, her experience points to some of the difficulties that may be encountered and anticipated in understanding the division theorem without a quotitive understanding of division.

Dividing A by D in the domain of the whole numbers is to determine the unique Q and R assured to exist by the division theorem. The common expression used for division of whole numbers, $A \div D = Q\text{rem}R$, however, is inexact (Gellert, Küstner, Hellwich, & Kästner, 1975). The use of the equality sign here can easily promote an impression that this "equation" is a well-formed arithmetic formula. Semantically, this equational form can be identified with the division theorem:

$$A \div D = Q\mathrm{rem}R \Leftrightarrow A = QD + R,\ 0 \leq R < D \qquad (*)$$

These forms of whole number division may lead the learner to believe that one can be derived from the other arithmetically in the same sense that, with rational numbers a, q, and $d \neq 0$, $a/d = q \Leftrightarrow a = qd$. This seems to have been the case with Andy and sheds light upon some of the difficulties that she experienced, in her second attempt above, with the integral approach to whole number division with the calculator. On her worksheet from the interview, she had written out the following "derivation"[3] in response to question set five:

$$\begin{aligned} A \div D &= Q\mathrm{rem}R \\ &= Q + R \\ &= D(Q + R) \end{aligned}$$

Andy substituted the predicate "rem" with the addition sign "+," which can be structurally interpreted as the equation: $A \div D = Q + R$. This interpretation could explain (1), her identification of the quotient with the sum of the quotient with the remainder. Andy then multiplied by D, which can be interpreted operationally as: D times $A \div D \Rightarrow D(Q + R)$. This interpretation could explain (2), why she multiplied the divisor with the sum of the quotient and remainder. In this latter case she noted that $24 \times (440 + 1)$ "doesn't work," apparently because the calculation failed to yield the dividend. In trying again, she implicitly used the division theorem, but note that she, like Armin above, expressed "the result" as: $A = DQ\mathrm{rem}R$, again interchanging "+" with "rem." This is consistent with (3), her expression of the product of the quotient and divisor as though it were the quotient.

The abstract relations between the quantities involved in whole number division, as defined and implied by the division theorem, were not obvious for the participants in this study. With question set four, the dividend was explicitly expressed in the form of the division theorem and participants were requested to determine quotients and remainders. A wide variety of responses were obtained of which only a brief summary will be presented here. Of the nineteen who were asked questions from this set, *only four* did not calculate the dividend *for any* of the questions they were asked. Rather, they, in one manner or another, adduced their answers directly from the form of the given expression. Of these four, only two, Stanley and Patricia, correctly identified the remainders and quotients for all the questions asked of them from this question set. Andy was one of the other two participants in this group. With the first part of the question set, "remainder and quotient of A divided by 6," Andy immediately offered 1 and 147, respectively. She also noted the remainder for A divided by 2 is "1, . . . because 2 is a factor of 6." However, when the interviewer asked for the remainder for 6 ×

147 + 2 divided by 2, Andy "guessed" the answer to be 2. Was Andy thinking of the second addend informally as "what was left over"? Was she unable to recognize and/or apply distributivity in resolving this problem? Did she forget, like Anabelle above, that the remainder must be less than the divisor? Unfortunately, whatever motivated her guess was not evident from the interview.

The other 15 participants, at one point or another, resorted to calculating the dividend and dividing, either via a calculator or manually, using long division. Two participants used a calculator. Both abandoned this approach for long division when, fixating on the fractional component of the rational quotient, they were unable to derive the remainder. A number of participants employed a variety of "even/odd number" heuristics. For instance, some participants, having calculated the dividend, noted it was either an even or odd number, and concluded that the remainder when dividing by 2 was 0 or 1, respectively. Others, noting that both addends of 6 × 147 + 2 were even numbers, concluded the remainder would be 0 when divided by 2. Nine participants, however, evaluated the dividend and essentially relied upon long division in addressing all questions from this set. Of those nine participants, Armin and Dana initially attempted alternative procedural approaches that also led to a rational quotient. Both, due to difficulties interpreting the fractional component, subsequently resorted to long division.

Some peculiar response patterns to this question set are particularly worth noting. Eight participants obtained a remainder for 6 × 147 + 1 divided by 6, by "canceling" the 6 from the first addend. Six of these then identified the remainder as 1, but Anabelle and Linda both added 147 + 1, incorrectly identifying the remainder as the sum of the quotient and remainder. No further questions from this set were asked of Anabelle. Linda, when asked for the quotient in this case, subsequently resorted to "doing it the long way": Calculating the dividend and performing long division. More notably in this regard, five of those six participants who had answered correctly with respect to division by 6 resorted to long division in considering the same dividend, 6 × 147 + 1, when divided by 2. Three of those five, Betty, Karen, and Lanny, calculated 6 × 147/2 and added the remaining addend to the result, again incorrectly identifying the quotient as the sum of the quotient and remainder.

Collectively, despite many other possible factors involved, these diverse results indicate the relation between whole number division with remainder and the division theorem, (*) noted above, was not evident for most of the participants in this study. Ironically, 15 of 19 participants, at some point in addressing question set four, resorted to a complex iterative application of the division theorem—long division.

Whole Number Division as the "Inverse" of Multiplication

Eight of the eleven participants who addressed this question tacitly considered the division theorem as the "inverse" of whole number division in the same sense that one would consider multiplication as the inverse of division in the domain of rational numbers. However, an inverse relation of whole number division with multiplication obtains only in the restricted sense of divisibility. Even with an understanding of divisibility, understanding whole number division, the division theorem, and its relation to multiplication is not always evident. For instance, Anita was one of the few participants that had exhibited a good conceptual understanding of divisibility in this study:

Interviewer: Okay. Um, now it's often been said that division is an inverse of multiplication. In what sense would you agree or disagree with that?

Anita: That division is an inverse of multiplication.

Interviewer: Uh huh.

Anita: Well if you just, by definition sense, multiplication is how many groups of something there is, and division is how many groups can go in a number, so you're basically just going backwards. Like um, it's like addition and subtraction, like $3 + 4$ is 7, $7 - 4 = 3$; multiplication 3×4 is 12, 12 divided by 4 is 3.

Interviewer: Alright. Um, how about division with remainder? Is there any way in which you see that as being an inverse to multiplication?

Anita: [*Pause*] Well whenever you're multiplying something, you always get a whole number, you never get a, not, not a whole number, but you always, well you always have a number, you never have a remainder in multiplying. [. . .] an inverse of something is just the opposite. Once you have a remainder it makes it totally different from multiplication.

When Anita said that with multiplication, "you always get a whole number," she seems tempted to say that "you don't always get a whole number" when dividing, but doesn't (perhaps because she realized that such a response would not be germane in considering division with remainder). Rather, she notes, in contrast with whole number division, that "you never have a remainder in multiplying." The remainder introduces an additional term in whole number division, making it "totally different" from multiplication.

In contrast, Dana exhibited strong procedural orientations in her responses to questions throughout her interview. She was creative in extending and applying abstract symbolic procedures from one context to another, although most often inappropriately. For her, a kind of inverse relation between division with remainder and multiplication was evident. While acknowledging Anita's concern, she found an inverse relation between division with remainder and multiplication (with addition) in terms of the division theorem:

Interviewer: In what sense does it have any meaning for you that division and multiplication are inverses of each other?

Dana: Well, if um $3 \times 4 = 12$, and 12 divided by 3 is 4 . . .

Interviewer: Okay. How about division with remainder?

Dana: Um, [*pause*] well, you never really see that [a remainder] in multiplication, but like I said, you take the quotient times the divisor plus the remainder gives you the dividend, so you, in checking to make sure you've done it right, you're doing multiplication.

Dana's understanding of the division theorem was more a matter of inverting a division procedure, presumably for checking the results of long division, rather than as a conceptual basis for understanding whole number division. It is somewhat ironic—if not confused—that the structural relation defining whole number division would be taken as it's own inverse.

Anita's concern and Dana's tendency to consider the division theorem as the inverse of division, or not, are both connected with the fact that the division theorem is based on a structural relationship involving both multiplication and addition. Their differences may be resolved if one considers multiplication (and addition) and division (and subtraction) more generally in terms of composition and decomposition, respectively. In this sense, the whole number dividend A can be interpreted, "inversely," either as the composition of $QD + R$, or as a decomposition into $QD + R$.

DISCUSSION

Although the focus of this study has been with preservice teachers' abilities to think at an abstract level about arithmetic division, the results should not be taken to necessarily reflect the understanding learners may have of division problems framed in more familiar computational or concrete situational contexts. Nevertheless, in this study it is evident, in the abstract contexts presented here, that participants encountered a wide diversity of difficulties when presented with novel problem tasks. Many of these difficulties surfaced or were indicated through inappropriate extensions and adaptations of more common and familiar procedures and concepts. Overall, the study indicates that preservice teacher content knowledge of arithmetic division, and other associated arithmetic procedures and concepts, can be somewhat rudimentary and fragile in more abstract contexts. Most of the preservice teachers participating in this study exhibited procedural orientations in their understandings of divisibility (Zazkis & Campbell, 1996a). This has been the case, more generally, with respect to division as well. These findings are in accord with previous studies, which have found preservice teachers' content knowledge to be procedurally orientated and conceptually "fragmented" (Ball, 1990), and "sparsely connected" (Simon, 1993).

Coming to Terms with Division

In this study, preservice teachers' understanding of arithmetic division in abstract contexts appears to be compromised by a lack of discernment regarding the numerical domains, symbolic expressions, and definitions involved. Such difficulties may be caused or exacerbated by contextual ambiguities of abstract terms such as *division, dividend, divisor, quotient*, and *remainder*. Question sets two, three, and four proved to be effective for uncovering these kinds of difficulties. In some cases, especially with question set two, it seems learners' difficulties were rooted well beneath such considerations. For instance, some problems seemed grounded in basic misconceptions regarding additive and multiplicative structure, others in basic calculation errors.

It appears that many obstacles learners confront in understanding arithmetic in their earlier years are left unresolved into adulthood. Given the immense number of obstacles students may encounter in learning arithmetic to adulthood, the potential permutations of unresolved obstacles to understanding arithmetic offer a plausible account for the wide diversity of difficulties encountered in this study. If so, this would imply that many, if not most, research findings that have focused on children's content knowledge are germane to preservice teachers' content knowledge as well. As a case in point, in the more abstract contexts of symbolic computation it is well known that middle school learners' understandings of connections between whole, fractional, and decimal numbers can be quite problematic (e.g., Mack, 1995; Markovits & Sowder, 1991; Resnick et al., 1989). Greer (1987), quoting Hiebert and Wearne (1983), notes that learners "see little connection between the meaning of a decimal and a whole number or a fraction." For preservice teachers in this study, this seems especially true regarding what have often been referred to by educators and researchers alike as "different forms of expression for remainders" (e.g., 10R1, 10½, or 10.5).

In particular, a variety of difficulties arose for participants in relating "decimal remainders" (the fractional component of the rational quotient) to whole number remainders. Greer (1987) has noted many children have difficulty understanding decimals. He cites Carpenter, Corbitt, Kepner, Lindquist, and Reys (1981) regarding a conception whereby a decimal number is interpreted as two integers separated by the decimal point. Resnick and colleagues (1989), expanding upon the work of Sackur-Grisvard and Leonard (1985), have investigated this phenomenon in detail and refer to it as the "whole number rule." It appears to have been a conception similar to this that Jennifer was standing under when she fell into a distrust of her calculator. Tara could not relate decimals to fractions at all.

A number of participants in this study, such as Anita and Jennifer, attempted to obtain a whole number remainder by first transforming the fractional component of the rational quotient from a decimal to a fraction. As pointed out above, some of these transformations can be friendlier than others, depending on the divisibility properties of the whole number remainder and divisor with respect to powers of 10. Even if successful in transforming from the decimal form to a fraction, connections between the remainder and numerator and the divisor and denominator may still not be evident to some learners.

Considered from the other direction (to obtain a rational quotient from a whole number division), the division theorem offers an effective starting point from which to discuss these important relations: $A = QD + R = QD + R(D/D) = D(Q + R/D) = Dq$. Although this particular task was not included *per se* in this study, this "simple" derivation highlights the central role of arithmetic division in making some of the fundamental procedural and conceptual shifts involved in numerical transitions from whole numbers to fractions to rational numbers. In particular, it helps to reveal the fundamental relations between the whole number quotient, divisor, and remainder with a rational quotient (namely, $Q + R/D = q$). It also serves to highlight some other important arithmetic notions such as multiplying and dividing a value by the same quantity and distributivity (Campbell & Zazkis, 1994). Some participants encountered various aspects of this derivation in question set four, whereas the last step is implicit in question set three.

Whole Number and Rational Number Division

Overall, the most important result of this investigation for mathematics teacher education is that it should not be assumed preservice teachers have an adequate understanding of the abstract similarities and differences between elementary whole number and rational number division. Given further research, some of the observations made in this study may prove to be more systematic and widespread than others. A possible case in point is the phenomena of confusing the quotient of a whole number division with the remainder. In this case, the problem seems rooted in the fact that whole number division, in contrast with rational number division, yields an answer with not one, but two components (the quotient and the remainder). Indeed, whole number division is unique in this regard, not only with respect to rational number division, but with respect to the other basic arithmetic operations as well. This may present a conflict with some intuitive or operational propensity for learners to expect a singular result from an arithmetic operation.

Interchanging referents for the whole number quotient and remainder have manifested themselves here in the novel context of divisibility and prime decomposition. In more familiar contexts it seems more likely that learners would consider the quotient as "the answer." The remainder, on the other hand, seems more likely to be considered not as part of a singular answer as it would be in rational number division, but rather, as that which is "left over," or "not included." In such cases only the quotient would be the answer. These notions may be accentuated by the common practice of referring to the answer of a whole number division: for example, 21 divided by 2, as "10, remainder 1," rather than as "*quotient* 10, remainder 1." Aversion to explicitly identifying a quotient in lieu of more comforting and tangible terms such as "the answer" is understandable. The term "quotient" can be vague and confusing due to its having different formal meanings with regard to whole number and rational number division, compounded by different conceptual meanings with regard to partitive or quotitive division. It is all the more esoteric and intimidating to the uninitiated as it is a term with no common usage in natural language.

Further in this regard, informal notions regarding whole number division may also be accentuated by the potentially misleading notation "A ÷ D = QremR." As we have seen above, this notation may be implicated in a variety of computational errors when taken as a standard arithmetic equation. When whole number division is considered in context of the division theorem, Andy and many others in this study illustrate that errors such as identifying the quotient as the product of the quotient and divisor, interchanging "rem" for "+," and adding the remainder to the quotient, should not be unexpected.

An analogous case for rational number division would be to interpret "rem" as the decimal point of the rational quotient. In tandem with the whole number rule this ties into another case that appears to be systematic and widespread among learners: equating the fractional component of the rational quotient (in decimal form) with the remainder. Some learners may simply be over-generalizing the fact that a zero to the right of the decimal point of a rational quotient always corresponds to a whole number remainder of zero. A more sophisticated account of this case, from a partitive perspective like Jim's, would be to identify the fractional component of the rational quotient (that part of the remainder in a given share) with the whole remainder itself. From a partitive point of view, it is a matter of realizing, as did both Jim and Stanley, that the cumulative "remaining parts" in each share constitute the whole, undivided, remainder. That is, the fractional component of the rational quotient multiplied by a whole number divisor is equal to the whole number remainder.

Partitive and Quotitive Dispositions Toward Division

Although not originally intended as a specific focus of investigation in the design of this study, quotitive and partitive dispositions toward division appear to be implicated in some way in participant success and failure in many of the tasks performed in this study. On a positive note, it is evident there are structural similarities between the division theorem, thinking of whole number division quotitively, and a quotitive approach to determining a whole number remainder from a rational quotient. It seems reasonable to suggest that problems in understanding the division theorem and using a calculator to identify remainders can be alleviated by a quotitive disposition toward whole number division. The results of this study, however, should not be taken as offering conclusive empirical evidence for this conjecture. Indeed, some caution seems warranted in that Simon (1993) observed no such correlation in his study of prospective teachers' approaches to similar problems. The intention here is simply to note: (1) there are structural similarities involved; (2) these similarities appear to correlate, and may be connected in some way, with participants' success in solving the calculator and division theorem problems; and (3) that, given the structural similarities alone, emphasizing the role of quotitive division with respect to whole number division with remainder as defined by the division theorem has pedagogical merit and warrants further investigation.

A notable feature of the intuitive models of division as proposed by Fischbein and colleagues (1985) is that, aside from the divisor in the partitive model, which intuitively must be a whole number, all other values may be taken to be either whole or rational numbers. In consequence, it seems, prima facie, natural to associate the partitive model with whole number division (divisor must be a whole number) and the quotitive model with rational number division (divisor can be a rational number). With respect to whole number division, however, the partitive model is unproblematic in a formal sense only in the restrictive case of divisibility (division with zero remainder). There are troublesome complications in thinking of whole number division partitively (in accord with Fischbein and colleagues' model) when a nonzero remainder is involved. To share a nonzero remainder equally means to create a fractional amount—*but there are no fractions in whole number arithmetic*. One reason why division is not "closed" in the domain of whole numbers is because whole number arithmetic units are indivisible—they cannot be denominated.

Thus, although they are related in many ways, it is important to remember that whole number division and rational number division are formally distinct operations. The intuitive partitive model, because it implicitly assumes rational number division in allowing a rational quotient, conflates their fundamental differences. Aside from the special case of divisibility, as indicated by Jim's reasoning above, one must ultimately divide a formally indivisible whole number arithmetic unit if one is to fully consummate the intuitive partitive model of division. This problematic, if not incommensurable, aspect of partitive division seems to lurk in the background of the fractional approach to the calculator problem of question set three and perhaps, to some extent, question sets four and five as well. No doubt there are other subtleties involved between intuitive models and formal aspects of division. Exactly how intuitive partitive and quotitive models of division may be implicated in learners' understandings of divisibility, whole number division, denomination, and rational number division warrants further investigation.

CONCLUDING REMARKS

This study has identified and described, from an empirical perspective, a variety of linguistic, procedural, and conceptual phenomena associated with preservice teachers' understandings of arithmetic division within abstract contexts. These phenomena all seem, in one way or another, related to: (1) an unfamiliarity of the connections between whole, fractional, and rational numbers and their symbolic expressions; (2) an overdependence on interpreting formal referents of arithmetic division using informal language; (3) a propensity to overgeneralize and misapply familiar procedures; and (4) a lack of discernment between whole and rational number arithmetic and a subsequent vulnerability to procedural and conceptual differences and referential ambiguities concerning the formal terms involved.

Theoretically, this study has highlighted a variety of linguistic, procedural, and conceptual distinctions that could, to some extent or another, be thematically involved in a more "ideal" understanding arithmetic division such as: (1) whole numbers and non-negative rational numbers; (2) indivisible and divisible units; (3) whole number division with remainder and rational number division; (4) integral and fractional approaches to division with remainder using calculators; and (5) quotitive and partitive division. Whole numbers, indivisible units, and division with remainder are clearly deeply related, as are non-negative rational numbers, divisible units, and rational number division. Quotitive division seems to have a natural, although not exclusive, affinity with the former group as does partitive division, although again not exclusively, with the latter group. The extent to which learners can thematize connections between these various formal and intuitive aspects of division along these lines remains to be more fully determined.

APPENDIX

Whole Number Division

The formal basis for whole number division is a fundamental theorem of number theory known as *the division algorithm*. According to this theorem, for any whole numbers, A and (nonzero) D, referred to as the dividend and divisor respectively, there exists a *unique* whole number quotient, Q, and a remainder, R, where $0 \leq R < D$, such that $A = QD + R$. Most learners are implicitly familiar with the division theorem, insofar as it is used iteratively in long division and as a means of "checking" the results thereof. Whole number division, however, is often expressed as $A \div D = Q\text{rem}R$. Note that this "equation" is imprecisely formulated and *does not* conform to standard arithmetic meaning and usage. Some learners may be prone to operating on this common expression of whole number division as if it were a well-formed arithmetic formula.

Divisibility

In the special case where $R = 0$, terms used in whole number division also refer, polysemously, to *divisibility*. For instance, in whole number division, "with remainder," any (nonzero) whole number can be designated as a divisor, D, of any whole number designated as a dividend, A. However, if divisibility is what is intended, and the remainder is nonzero, D is neither a divisor of A, nor can it be said to divide A. See Zazkis (Chapter 4, this volume) for other ways of defining and thinking about divisibility.

Rational Number Division

In the domain of rational numbers, division may be more intuitive, but conceptually it is more involved. It depends on the fact that for any (nonzero) rational number, *d*, there exists a unique multiplicative inverse, or reciprocal, d^{-1}. In dividing a dividend, a, by a divisor, d, the quotient, q, is defined as the product, ad^{-1}. The quotient can be expressed in a number of equivalent ways: $ad^{-1} = a/d = a \div d = q$; or implicitly, in an equation more closely related to the division theorem and divisibility, $a = qd$. *There is no such thing as a remainder in rational number division.* Instead, as is characteristic of rational numbers, the quotient has a decimal, or *fractional component* (possibly zero) in addition to a whole number, or *integral component.* The relationship between remainders and fractional components is not always evident to learners. Indeed, it may be tempting for some learners to simply identify whole number remainders with fractional components of rational number quotients. Such imprecise designations can present many difficulties to learners in their attempts to understand division.

Quantitatively Equivalent Expressions

The same number can be expressed in a wide variety of different forms and representations. The variety of intended meanings and expressive possibilities for particular numbers are virtually endless. More importantly, considering numbers from such different points of view can foreground some aspects of meaning from others. For instance, and of particular interest in this study, the prime decomposition of a given dividend can foreground that number's multiplicative structure, whereas the linear form of the division theorem can foreground its modular structure. Some of the analytic and semantic implications of these forms of numerical expression are not always evident to learners.

Whole Number Division Using Calculators

At a user level, division with calculators involves the domain of rational numbers. There are two basic approaches to determining the remainder from a rational quotient obtained from a whole number division using a calculator (Simon, 1993). The *integral approach* dispenses with the fractional component of the quotient and involves the difference between the product of the integral component of the quotient and the divisor, and the dividend. The *fractional approach* dispenses with the integral component of the quotient and involves multiplying the fractional component of the quotient with the divisor.

Note that once the integral component of the quotient is determined, the integral approach ostensibly involves calculations with whole numbers, whereas the fractional approach continues to conflate the use of whole and rational numbers. Furthermore, if the fractional component of the quotient is truncated, as is most often the case due to limitations in the precision of the calculator, the resultant "remainder" has to be "rounded up" to the next whole number, in order to compensate for the discrepancy.

Partitive and Quotitive Division

An important interpretive shift in story problem contexts for learners concerns *partitive* and *quotitive* models of arithmetic division (e.g., Fischbein et al., 1985). Freudenthal (1983) has considered partitive and quotitive interpretations in terms of the phenomenological acts presupposed in making them. In this chapter, dispositions toward partitive and quotitive division are characterized in a more general and abstract sense. A partitive disposition toward division is evident when the (partitive) quotient is intended as, or understood to be, the value of the resultant parts. Intuitively, the number of parts, the divisor, is taken to be a nonzero whole number (Fischbein et al., 1985). A quotitive disposition is evident when the quotient is intended as, or understood to be, the number of times the divisor goes into the dividend. Thus, the (quotitive) quotient can be seen as a measure of the dividend in units of the divisor.

NOTES

1. A preliminary report of this research was presented at the 20th International Conference for Psychology of Mathematics Education, in Valencia, Spain. This work was made possible in part through support from the Department of Education, University of California, Irvine, Grant No. 410–93–1129, and Fellowship No. 752–94–1246 from the Social Sciences and Humanities Research Council of Canada.

2. In this chapter, capital letters are used to designate whole number (non-negative integer) variables and small letters are used to designate (non-negative) rational number.

3. The actual letters Andy used on her worksheet, A, B, C, and D, have been substituted with A, D, Q, and R, respectively, to correspond with (*).

REFERENCES

Ball, D. (1990). Prospective elementary and secondary teachers' understanding of division. *Journal for Research in Mathematics Education, 21*, 132–144.

Campbell, S., & Zazkis, R. (1994, November). Distributive flaws: Latent conceptual gaps in preservice teachers' understanding of the property relating multiplication to addition. In D. Kirshner (Ed.), *Proceedings of the 16th Annual Meeting of the North American Chapter of the International Group for the Psychology of Mathematics Education* (Vol. 2, pp. 268–274). Baton Rouge: Louisiana State University.

Campbell, S. R. (1996). On preservice teachers' understandings of division with remainder. In L. Puig & A. Gutiérrez (Eds.), *Proceedings of the 20th International Conference for Psychology of Mathematics Education* (Vol. 2, pp. 177–184). Valencia, Spain: Universitat de València.

Fischbein, E., Deri, M., Nello, M. S., & Marino, M. S. (1985). The role of implicit models in solving verbal problems in multiplication and division. *Journal for Research in Mathematics Education, 16*, 3–17.

Freudenthal, H. (1983). *The didactical phenomenology of mathematical structures*. Dordrecht, The Netherlands: Reidel.

Gellert, W., Küstner, H., Hellwich, M., & Kästner, H. (Eds.) (1975). *The VNR concise encyclopedia of mathematics*. New York: Van Nostrand Reinhold.

Greer, B. (1987). Nonconservation of multiplication and division involving decimals. *Journal for Research in Mathematics Education, 18*, 37–45.

Mack, N. (1995). Confounding whole-number and fraction concepts when building on informal knowledge. *Journal for Research in Mathematics Education, 26*, 422–441.

Markovits, Z., & Sowder, J. T. (1991). Students' understanding of the relationship between fractions and decimals. *Focus on Learning Problems in Mathematics, 13*(1), 3–11.

Matz, M. (1982). Towards a process model for high school algebra errors. In D. Sleeman & J. S. Brown (Eds.), *Intelligent tutoring systems* (pp. 25–50). London: Academic Press.

National Council of Teachers of Mathematics. (1989). *Curriculum and evaluation standards for school mathematics*. Reston, VA: Author.

National Council of Teachers of Mathematics (1991). *Professional standards for teaching mathematics*. Reston, VA: Author.

National Council of Teachers of Mathematics. (2000). *Principles and standards for school mathematics*. Reston, VA: Author.

Resnick, L. B., Nesher, P., Leonard, F., Magone, M., Omanson, S., & Peled, I. (1989). Conceptual bases of arithmetic errors: The case of decimal fractions. *Journal for Research in Mathematics Education, 20*, 8–27.

Simon, M. (1993). Prospective elementary teachers' knowledge of division. *Journal for Research in Mathematics Education, 24*, 233–254.

Vergnaud, G. (1996). The theory of conceptual fields. In L. P. Steffe, P. Nesher, P. Cobb, G. A. Goldin, & B. Greer (Eds.), *Theories of mathematical learning* (pp. 219–239). Mahwah, NJ: Erlbaum.

Wheeler, M. M., & Feghali, I. (1983). Much ado about nothing: Preservice elementary school teachers' concept of zero. *Journal for Research in Mathematics Education, 14*, 147–155.

Zazkis, R., & Campbell, S. R. (1996a). Divisibility and multiplicative structure of natural numbers: Preservice teachers' understanding. *Journal for Research in Mathematics Education, 27*, 540–563.

Zazkis, R., & Campbell, S. R. (1996b). Prime decomposition: Understanding uniqueness. *Journal of Mathematical Behavior, 15*, 207–218.

3

Conceptions of Divisibility: Success and Understanding

Anne Brown, Karen Thomas, and Georgia Tolias

This study[1] contributes to the research literature concerning prospective elementary teachers' understandings of basic concepts of elementary number theory. Our analysis of how college students apply their conceptions of multiplication and division to problem situations involving divisibility is structured by a theoretical framework that combines APOS theory with a stage model adapted from Piaget's work in *Success and Understanding* (1978). We are interested in the individual's ability to progress from responding in ways that are primarily action-oriented, with little awareness of the underlying mathematical concepts, to responding with inferential reasoning based explicitly on an understanding of mathematical operations and properties.

In examining prospective elementary teachers' understandings of elementary number theory, we focus on the individual's ability to guide her or his arithmetic actions and reasoning through an understanding of the multiplicative structure of the set of natural numbers. We use the following characterization of multiplicative structure provided by Freudenthal:

> The multiplicative structure of **N** is the whole of the relations $a \times b = c$, possibly also expressed as $c \div b = a$, complemented by $a \times b \times c = d$, $a \times b = d \div c$, and all one can think about in this context. At a higher level, it includes experience, and at a still higher level formulated knowledge, of such properties as commutativity, associativity, distributivity, and equivalence of $a \times b = c$ and $c \div a = b$, many more properties of this kind. (1983, pp. 112–113)

Attention to the multiplicative structure of natural numbers occurs at all levels of mathematics—from its introduction in elementary school to the frontiers of mathematical research. The study of elementary number theory is a particularly appropriate subject for future teachers because applying their knowledge of multiplicative structure in this context provides valuable opportunities to enrich their understanding of the properties of multiplication and division.

To take full advantage of the concept of multiplicative structure, the individual must have experience with the representation of natural numbers as the product of primes. This includes constructing prime factorizations, performing arithmetic on prime factorizations, and using the structure embedded in the factorizations to recognize and justify divisibility relationships. The elementary nature of the process of constructing prime factorizations for small numbers may suggest to some that the use of these factorizations as a structural tool for problem solving is equally elementary. To the contrary, recent research on the learning of number theory carried out by Zazkis and Campbell (1996a, 1996b) indicates that many prospective elementary teachers have difficulties working with this representation, in that fundamental properties and relations are often not easily recognized or inferred from structural descriptions.

In our analysis of prospective elementary teachers' use of the concept of multiplicative structure, we employ a combination of two theoretical approaches. Our primary theoretical tool is APOS theory, which is described below. In addition, we have found it useful to draw upon the work of Piaget that is presented in his book *Success and Understanding* (1978). We were led in this direction by our observation that our students often deal with number theoretic tasks without consciously using their knowledge of multiplicative structure. For example, they might choose to actually perform computations when reasoning about computations would suffice. This dependence on know-how rather than reasoning struck us as reminiscent of what Piaget observed in a series of experiments involving children's success with physical tasks. His subjects were often successful with the tasks while remaining unaware of the role of general mechanisms whose interplay was responsible for the effectiveness of their actions. Moreover, he described a progression through which one might move from successful action to understanding that we found helpful, in analogy, for analyzing our subjects' work on divisibility tasks. In the next section, we provide additional detail on Piaget's work in this area and describe more precisely its relevance to the present study.

BACKGROUND

In this section, we lay the theoretical foundations for the design of our study, including a description of *APOS theory*. Our study builds on a previous study of divisibility conceptions using APOS theory that was carried out by Zazkis and Campbell (1996a), which is described more fully below. Further, we describe the relevance of Piaget's results from *Success and Understanding* to the present study.

Our theoretical perspective on mathematical knowledge is as follows: "An individual's mathematical knowledge is her or his tendency to respond to perceived mathematical problem situations by reflecting on problems and their solutions in a social context and by constructing or reconstructing mathematical actions, processes, and objects and organizing them in schemas to use in dealing with the situations" (Asiala et al., 1996, p. 7). Each of the terms—action, process, object, and schema—defines a mental construction one might make in order to understand a mathematical idea. The theory holds that one starts from an action conception of a mathematical idea, and moves ultimately (but perhaps not linearly) to constructing a schema for an area of mathematical content. What follows is a general description of each of the theoretical constructs included in the APOS perspective, followed by an example. The reader may also wish to consult Asiala and colleagues (1996) for a more complete introduction to the theory.

An *action* is any transformation of objects to obtain other objects. It is perceived by the individual as being at least somewhat externally directed, as it has the characteristic that at each step, the next step is triggered by what has come before. An individual is said to have an *action conception* of a given concept if her or his depth of understanding is limited to performing actions relative to that concept. Someone with a more advanced understanding of a concept may well perform actions when appropriate, but is considered to have moved beyond an action conception if he or she is not limited to performing actions.

When an action is repeated, and the individual reflects upon it, it may be interiorized to a mental *process*. In contrast to actions, processes are perceived as being internal to, and under the control of, the individual. An individual is said to have a *process conception* of a given concept if the individual's depth of understanding is limited to thinking about the idea as a process. An important characteristic of a process conception is that the individual is able to describe, or reflect upon, the steps of the transformation without actually performing those steps. New processes may be constructed from existing ones through reversal, or through coordination with other processes.

If an individual reflects on operations applied to a particular process, becomes aware of the process as a totality, realizes that transformations can act on it, and is able to actually construct such transformations, then we say the individual has constructed this process as a cognitive *object*. In this case, we say that the process has been *encapsulated* as an object. An individual is said to have an *object conception* of a mathematical concept when he or she becomes able to treat that idea or concept as a cognitive object, including being able to perform actions on the object and reason about properties of the object. Such an individual is also able to *de-encapsulate* an object back into the process from which it came in order to use its properties.

A *schema* for a certain piece of mathematics is an individual's collection of actions, processes, objects, and other schema that is linked to form a coherent framework in the individual's mind, and that may be brought to bear upon a problem situation involving that area of mathematics. The individual may or may not be conscious of the framework that gives rise to the links. An important and defining characteristic of the coherence of the framework is its use in deciding what is in the scope of the schema and what is not, knowing what the capabilities of the schema are, and being able to recognize situations in which the schema can be applied. A schema can be *thematized* to become a cognitive object to which actions and processes can be applied.[2]

As an example, we consider the concept of function. First, an individual who is unable to interpret a situation as a function unless he or she has a single formula, with explicit instructions for computing values, is restricted to an action conception of a function. An individual who has a process conception of function can think of a function in terms of accepting inputs, manipulating them in some way, and producing outputs. Someone with a process conception of function can invert and compose functions when necessary. Indications that an individual might have an object conception of function include being able to form sets of functions and being able to perform arithmetic operations on functions. Indications that an individual might have thematized her or his function schema include the ability to construct examples of functions that fit specified conditions.

When a mathematical topic is analyzed using APOS theory, a major goal is often to produce a *genetic decomposition* for the topic, that is, a description of specific mental constructions that a learner might make to develop her or his understanding of the concepts. Genetic decompositions are used to inform the design of instruction and are subject to revision following an assessment of the effectiveness of the instruction. This approach has been used successfully to examine college students' understanding of a variety of mathematical topics including calculus, statistics, mathematical induction, and abstract algebra (see, e.g., Asiala et al., 1996; Clark et al., 1997; Clark et al., 2000; Dubinsky & McDonald, in press, for more information). Prior to presenting our APOS analysis, we describe Zazkis and Campbell's earlier application of APOS theory to the understanding of divisibility concepts.

Summary of the Zazkis and Campbell Study

Zazkis and Campbell (1996a) gave an analysis of student responses to a collection of questions about divisibility and the multiplicative structure of natural numbers. The questions were selected from clinical interviews, addressing a variety of elementary number theory topics that were conducted with twenty-one prospective elementary teachers. Their analysis and interpretation of the interview data were grounded in APOS theory, and used a genetic decomposition for divisibility, which we briefly describe in terms defined above. In their view, the construction of the concept of divisibility starts with consideration of specific small numbers acting as divisors. An individual with an *action concep-*

tion of divisibility needs to actively perform division and make the decision about divisibility, after the fact, by considering the result. Later, this action may be interiorized as a mental process in which the division is intended but not actually performed. That is, the individual with a *process conception of divisibility* understands that a division procedure can be used to determine the divisibility decision, but does not always find it necessary to actually carry out the division to make the decision. When the concept of divisibility is seen as a bivalent "yes or no" property of whole numbers independent of the procedural aspects of division, the individual has encapsulated divisibility as a cognitive object, reaching an *object conception of divisibility*. Someone with an object conception of divisibility is able to perform actions on the property of divisibility; one such action is that of comparing hypotheses and conclusions to infer, for example, that $b + c$ is divisible by a when both b and c are divisible by a.

Aside from the genetic decomposition of divisibility given above, there are two important results of Zazkis and Campbell's (1996a) study, from our point of view. First, while students often narrowly associate divisibility with the action of division, their data suggest that to construct a broadly useful schema for divisibility, cognitive links with the concepts of multiplication, factors, and prime factorization are essential. Second, their study points out the need for additional emphasis in instruction on the central roles played by both division and multiplication in considering questions of divisibility, and in particular on the inverse relation between these operations (see Campbell, Chapter 2, this volume).

Piaget's Studies on Success and Understanding

In this section, we summarize our understanding of Piaget's (1978) studies of children's successes with certain physical tasks. As Sierpinska (1994) points out, Piaget's use of the word "understanding" in this work connotes both explicative and implicative understanding—one who understands can explain how and why actions are successful as well as predict whether or not a planned action can be successful.

In Piaget's work, we see that success with physical tasks can occur with very little understanding of why one's actions work, or how to structure actions to be consistently successful. Briefly, Piaget described three stages in moving from success to understanding: from **Stage I**, performing actions successfully with little awareness of general mechanisms at work in the task (knowing by doing); to **Stage II**, where actions and conceptualization have reciprocal effects on one another; and finally, to **Stage III**, when actions are consciously guided by and reasoned about through applying one's conceptualization of the task.[3]

A brief example from Piaget's work might clarify the ideas here. In one experiment, Piaget interacted with children as they tried to guide a boat, floating in a large bowl, toward a target. To become consistently successful with the task, one needed to comprehend the coordination of the sail and the rudder of the boat, although individuals could sometimes successfully guide the boat to the target without being able to explain the coordination. There were some participants who ignored the roles of the sail and the rudder, others who experimented with their coordination, and still others who were able to infer the results of coordinating the sail and rudder without having to experiment. This experiment was one in a series whose analysis contributed to the construction of the stage model presented in Piaget's work.

More generally, Stage I is typified by a display of a high level of practical skill that brings success in a narrow range of tasks, or in a limited context. Any conceptualization that takes place is focused on the external results of actions. Only after Stage I is there an interplay of action and conceptualization. During Stage II, reflection on actions builds understanding. But also, actions are affected by one's understanding, in that the results of actions are interpreted through the lens of what is understood, and what is understood suggests which actions to take. There is a growing awareness of the mechanisms at play, and how they are coordinated. When hypotheses are formed, they are tested through immediate application, rather than being examined through reasoning. In Stage III, the individual can reason about the mechanisms; issues originally thought of in terms of cause and effect are transformed through reconstruction to a process of inferential reasoning. Rather than just seeing how mechanisms are coordinated, there is an understanding of why they can be coordinated. The role of general mechanisms in one's success with the task is understood conceptually through reasoning, not only through action. Consequently, one can predict whether or not a sequence of particular actions will contribute to completing the task. This is what is meant by success with understanding: actions related to the task are guided by the individual's conceptualization of the task.

THEORETICAL ANALYSIS: APPLYING A SCHEMA FOR MULTIPLICATIVE STRUCTURE

In this section, we present our preliminary theoretical analysis of the application of one's schema for multiplicative structure to problems that involve reasoning about divisibility. This model will be used to analyze student responses to interview tasks that require applying and coordinating a variety of actions or processes relating to elementary concepts such as the divisibility relation, the concepts of divisor and multiple, and prime factorization.

In view of Freudenthal's (1983) characterization of multiplicative structure of the natural numbers, we propose that the components of an individual's schema for multiplicative structure might include schemas for the arithmetic operations (particularly the processes of multiplication and division, and the fundamental reverse[4] relationship between them), schemas for their properties (such as commutativity, associativity, and distributivity), and a schema for the unique factorization of natural numbers as the product of primes. Since these mathematical topics, in their basic form, are part of the content of the elementary and middle school curriculum, we assume that the components of a college student's schema for multiplicative structure have been constructed largely through her or his earlier mathematical experiences.

It is important to note that we are making no assumptions about whether or not any individual in our study has a divisibility schema. We agree with Zazkis and Campbell that divisibility is a "very complex cognitive structure" (1996a, p. 561). One goal of the present study is to learn more about the divisibility schema so that eventually we will be able to describe its development, possibly using methods such as those presented by Clark and colleagues (1997). In particular, in order to make progress in this endeavor we must understand more fully the ways in which a student's schema for multiplicative structure is applied in reasoning about divisibility.

To analyze data concerning the application of a schema for multiplicative structure, we need to consider situations in which an individual appears unable to handle situations in which we would expect the schema for multiplicative structure to be applied. In some cases, it may be that some components of the schema are not adequately constructed; APOS theory can be used to describe the constructions and analyze the nature of the weaknesses of these constructions. In other cases, the issue may be that the individual is unaware of the implications of the knowledge they have about multiplication and division. To analyze the metacognitive issue of the individual's ability to access and consider possible uses of her or his knowledge, we use the following stage model.

Drawing upon Piaget's (1978) three stages for moving from success to understanding on concrete physical tasks, we propose that prospective elementary teachers have a tendency to respond to divisibility tasks in one of the following ways:

1. by performing actions that reveal little awareness of the role of multiplicative structure, or
2. by making hypotheses that are tested through actions that reveal some awareness of the relevance of multiplicative structure, or
3. by making conscious and deliberate inferences about multiplicative structure that need not be tested through actions.

In more detail, our theoretical model is as follows:

Stage I. Initially, college students may use an action-oriented, *ad hoc* approach to questions about divisibility. The questions are typically understood to require examination of the results of arithmetic actions, particularly division. Attention to surface features captured in mathematical notation may substitute for reasoning about the operations responsible for those features. While individuals might be successful in answering questions about divisibility in a limited context (e.g., for small numbers whose multiplicative structure is familiar), they might not grasp the role of general mechanisms, such as the structural properties of multiplication and division, and the reverse relation between multiplication and division. They may respond in a way that is sufficient in dealing with a single case or a limited context, but which cannot be generalized.

Stage II. In this transitional stage, an application of one's schema to the task concentrates on structuring immediate actions and considering their results. This may take the form of a process of "trial and error," in which the result of an action leads to a conscious adjustment or regulation of a future action. However, the individual still cannot make inferences about the success or failure of actions without actually carrying them out. A growing awareness of how one's knowledge of multiplicative structure can or should apply might result in the individual reconstructing her or his schema for multiplicative structure to accommodate a broader variety of situations.

Stage III. Work at this stage is typified by conscious and deliberate applications of an individual's knowledge of multiplicative structure, including inferential reasoning about divisibility concepts that draws upon the individual's awareness of the schema's coherent underlying structure. The consideration of one's knowledge in relation to the task concentrates on explaining why certain actions succeed, and why others do not. Work in this stage includes the ability to make predictions of the success of future actions without direct experimentation.

We use this preliminary theoretical analysis to interpret student responses to a variety of divisibility tasks, and then evaluate its usefulness as an analytical tool in the final section of this chapter.

METHODOLOGY

Participants

The participants in this study were students enrolled in mathematics courses for prospective elementary school teachers taught by two of the authors at two regional universities in the midwestern United States. A combined total of seventy-three students were enrolled in these courses; of these, ten responded to a call for volunteers to be interviewed at the end of the course. There were eight women and two men interviewed. The final course grades earned by the interviewees were as follows: three As, five Bs, and two Cs.

Nature of the Course

This mathematics course focuses on the foundations of arithmetic. It uses an instructional approach based generally on a constructivist theory of learning. Students are expected to assume an active role in their mathematics learning by working in small groups on specific tasks while explaining their solutions to their group members, making sense of others' solutions, and challenging flawed explanations. Although the classes taught by the two authors followed this general instructional approach, the courses differed in the nature of the curriculum materials. One class used a standard mathematics textbook for elementary teachers that was supplemented by occasional class activity sheets, while the other class used only class activity sheets that were distributed to students on a daily basis—these activity sheets were generally not the same as those used in the other course. It is important to note that the design of the materials used in these courses was not informed, consciously, by APOS theory, or any other specific theoretical perspective.

Data Collection

The participants were interviewed individually either during the last week of classes or the final exam week of the semester in which they were enrolled in the course. The interviews were conducted by two of the authors of this study and they lasted from 60 to 90 minutes. There were three factors that influenced the design of the interview tasks: various aspects of the Zazkis and Campbell study, our collective experiences teaching these topics to prospective teachers, and our view of mathematical knowledge expressed in terms of APOS theory. The interviewers followed a protocol that outlined possible prompts and follow-up questions, but allowed for further probing at the discretion of the interviewer. Participants were allowed to use calculators. The interviews were audiotaped and transcribed verbatim, and the students' written work was collected.

The Interview Questions

The data presented in this chapter represents a small portion of the responses obtained when the interviewees were presented with the following tasks and asked to describe their thinking processes. The list of questions is presented first; it is followed by a description of what we hoped to learn about students' conceptions from each question.

1. Consider the following list of numbers: 1, 3, 6, 7, 8, 12, 15, 18, 24, 39, 42, 48, 69, 96, 2400, 2401, 2412
   (a) The numbers in the list that are divisors of 24 are:
   (b) The numbers in the list that are multiples of 24 are:
   (c) The numbers in the list that are divisible by 24 are:
   (d) The numbers in the list that 24 is divisible by are:
2. Describe in your own words what is meant by a
   (a) divisor of a number
   (b) multiple of a number
3. Describe in your own words what it means to say "M is divisible by N."

4. Consider the number $M = 3^3 \times 5^2 \times 7$
   (a) Is M divisible by 7? Explain.
   (b) Is M divisible by 5? by 2? by 9? by 63? by 11? by 15? Explain.
   (c) Is $3^2 \times 5 \times 7^3$ a multiple of M? Explain.
   (d) Is $3^4 \times 5^5 \times 7^3 \times 13^{18}$ a multiple of M? Explain.
5. Find the smallest counting number that is a multiple of both 72 and 378. Explain.
6. A scientist starts two experiments at the same instant. In the first experiment, a measurement has to be made every 168 seconds, while in the second, a measurement has to be made every 108 seconds. After how many seconds will the scientist have to make two measurements at the same instant? Explain.
7. Find a pair of numbers, each smaller than 200, whose least common multiple is 200. Explain your answer. Find another pair, different from the first pair.

The purpose of Question 1 was to give participants an opportunity to activate their knowledge about divisibility concepts at a concrete level. This activity led into Questions 2 and 3, where they were asked to articulate the meaning of each of the fundamental terms as they understood them. Eventually, all subjects were asked to provide a general definition of each term, and to address any inconsistencies with their answers to Question 1.

Questions 4(a) and 4(b) were designed by Zazkis and Campbell for their study. Our purpose for including these questions, as well as Questions 4(c) and 4(d), was to observe the extent to which knowledge of multiplicative structure influenced responses to tasks in which the numbers are expressed in prime-factored form. This also gave us the opportunity to compare the participants' views of the fundamental concepts in this setting with the descriptions they provided in responding to Questions 2 and 3. We hoped for a fruitful comparison, since we expected the responses to Questions 2 and 3 to be based primarily on their experiences with divisibility relations for numbers expressed in decimal form.

Questions 5, 6, and 7 all relate to student conceptions of the least common multiple, although only the last question used the term explicitly. We were particularly interested in the strategies for computing the least common multiple that would be evoked by the situations in Questions 5 and 6, and in the participants' ability to handle reverse processes in Question 7.

DATA ANALYSIS

For the most part the data analysis is organized following the order of the questions asked. By presenting a series of brief case studies from our interviews, we illustrate a variety of ways in which an individual might draw upon her or his schema for multiplicative structure to solve problems involving divisibility concepts (in the excerpts presented, the names of the students have been changed to protect their identities, and both interviewers are designated by the symbol "I").

Included in most of the cases are preliminary genetic decompositions, expressed in terms of APOS theory, for the concepts under discussion. It should be understood that each genetic decomposition is based on our own experiences as learners and as teachers, as well as on preliminary analyses of all of the data collected during this study. The decompositions are not very detailed and are certainly subject to revision; they are intended to structure and clarify the analysis presented in each case study.

Interview Questions 1, 2, and 3 were intended to help students activate their knowledge of divisibility concepts at a concrete level, and then summarize their understandings by stating general conditions. When necessary, the interviewer prompted individuals to identify, in terms of a process, or in terms of conditions on the object, how decisions were reached for the numbers in the list. The issue of consistency across the responses was also raised by the interviewer. In particular, since the meaning of divisible by, multiple of, and divisor of describe the same multiplicative relationships in different ways (see Zazkis, Chapter 4, this volume), we would expect a conscious application of one's knowledge of multiplicative structure to result in a statement referring to the equivalence of parts 1(a) and 1(d) and parts 1(b) and 1(c). Indeed, the responses of five of the ten participants generally fit that description.

At the same time, the initial responses of all of our subjects revealed a tendency to strongly associate the concepts of divisibility and divisor with the action of dividing, even though the formal mathematical definitions used in the course express these ideas in terms of multiplicative relationships. It was also typical to associate the concept of multiple with multiplication or repeated addition, perhaps reflecting the explicit attention to generating sequences of consecutive multiples that took place during instruction.

An Action-Oriented Approach to Divisibility Tasks

The responses of one of the participants, Adam, were characterized by a concentration on the role of division and a consistent tendency to choose to perform procedures rather than making inferences. In this section, we analyze the nature and possible consequences of his approach.

When Adam considers the divisors of 24 (in Question 1[a]), he begins to calculate immediately, rather than verbalizing a condition that divisors of 24 must satisfy. In his calculations, he refers to the results of intended division or multiplication, to products being "too big" or "too small," and to possible divisors as "working" or not. For example:

Adam: Uh, number sense, I know that 24 will not go, or 7 won't go into 24 so I know 7 doesn't work, um, number sense for 8, I know that works 'cause

> 8 will go into 24 . . . because 24 divided by 8 is 3. Um, for 12, I know number sense because 12 is ½ of 24 so 24, or 12 works, um. For 15, um, I, I would do, I would just start, I know 15 times 2 is too big so, so 15 times 2 is too big that means nothing works 'cause 15 times 2 won't work.

At the end of Question 1, the interviewer asks him to revisit part (a) to give further explanation of how he was making his decisions. With some prompting he recalls what his original reasoning actually was. He says, in part, ". . . if I remember, I was taking 24 and dividing it by 3 and seeing if it would go evenly," meaning that the division will "leave me with a whole number." Note that even his discussion of his overall strategy includes reference to specific actions. At the same time, we see in his responses signs of progress toward interiorization of his actions related to divisors. For example, his observation that "nothing works 'cause 15 times 2 won't work" suggests that he may be in transition from action to process, in that he can imagine the results of the process without having to carry it out.

Adam's initial response to divisibility tasks tends to be to calculate rather than to anticipate, infer, or predict. The automatic nature of his response might present an obstacle to reflection, judging from his response to Question 2, when he was asked to explain, in his own words, what a divisor is:

Adam: A divisor, I'll write it first . . . so I can make sure I get it right.

I: Ok.

Adam: [*A pause, almost one minute in length*] See if I can write it. I know how to do it. Um.

I: So . . . just explain in your own words, when somebody says, tell me what a divisor of a number means or is, what would you say?

Adam: To me a divisor of a number is, I would say, it's a number that's divisible by, an—any numbers that, that are, any numbers that are divisible, or any numbers that can go into the numbers that I'm divising. . . .

The long pause in responding to a request to describe a divisor of a number in his own words, as well as the awkwardness of his eventual response, suggest that consciously formulating a general process or condition for a divisor is a challenge for him, and something he may not have done before, even though, as he puts it, he knows "how to do it."

When considering the multiples of 24 (in Question 1[b]), Adam is confused initially, and tries to recall a condition defining multiples, but cannot. At first, he includes the divisors of 24 among its multiples. In his return to action in order to consider the listed numbers that are larger than 24, he recalls a meaning for multiple:

Adam: I know 39 doesn't, doesn't work because for it to be a multiple it would have to be like 20—, for me, I find it has to be like 24 times 2 like would be the first multiple you could have, it could be the smallest one, next so I know that 39 and 42 don't work because they're both smaller than what 24 times 2 is, because 24 times 2 is 48 so then I know 48'll work.

Adam shows that, in this case, he is aware of the role of multiplication in generating multiples of 24, and he can identify nonmultiples in relation to the benchmarks formed by the multiples of 24. He continues:

Adam: Um, and then I would take 24 times 3, which is 72 so then I know that 69 doesn't work because it's too small and since there isn't a 72 [in the list given in the question] then I would go to my next number which I would take 24 times 4 which would be 98, so, [*works on calculator, pause*] no, 96, there we go, 24 times 4 is 96, so 96 would work 'cause then 20, 'cause 24 times 4 works which makes it a multiple [*inaudible*], um, to find, for 24 hundred I would take 24 hundred divided by 24 and see if, if 24 is a divisor of 24 hundred, then I know that 24 hundred is a multiple of 24.

Viewing multiples of 24 as numbers that have 24 as a divisor results in Adam's translation of the question to one that requires carrying out direct division:

Adam: So, take 24 hundred divided by 24 [*carries this out on his calculator*]. I find it does go evenly and it's 100, so that's 100, so that tells me that 24 hundred is a multiple, um, 24 hundred and one doesn't work because I don't think that goes evenly [*divides using the calculator*] and it doesn't, so I know 2401's not a multiple and then I do the same thing for 2412 and it doesn't go evenly so it's not a multiple.

Curiously, Adam does not find the expression "24 hundred" meaningful, but instead carries out division on his calculator to determine whether it is a multiple of 24. While he predicts that "24 hundred and one doesn't work," he finds it necessary to calculate to confirm. We interpret his behavior here as moving from Stage I to Stage II. Rather than simply calculating, he formulates a prediction and tests it with immediate action. At the same time, it is clear that he is not at Stage III on this task: his inability or reluctance to infer the result of division instead of actually dividing suggests that he may be unaware of the role of multiples of N as benchmarks in the process of division by N. That is, his earlier use of benchmarks may have been only an action, rather than a conscious part of a process for determining the result of one natural number being divided by another.

When the issue of a general definition of multiple is raised in Question 2, here is what Adam says (this immediately follows his attempt to define a divisor of a number, excerpted above):

> Adam: A multiple, um to me is um, it's, it's pretty much the op—, it's almost the opposite . . . it's um, it has to be numbers that are equal or bigger . . . than the original number and um, the, the, the orig—, the original number has to in turn be a divisor of the multiple or, or the bigger number.
>
> I: Good. Right.
>
> Adam: And it has to be able to divide into those numbers.

All mention of multiplication has disappeared here, which suggests that the procedure of generating multiples through multiplication is only an action for him; in fact, throughout the entire interview, he shows no tendency to view his actions as producing multiples endlessly. Moreover, the fact that he feels it is necessary to mention both the word "divisor" and the condition "it has to be able to divide into those numbers" provides further evidence that the concept of a divisor is not meaningful to him without a direct reference to an action.

In considering which numbers 24 is divisible by, Adam restates the issue immediately as a division problem, saying "I know that 24 has to be the numerator for this division-like problem and so any of the bigger numbers won't work because 24 over any of the bigger numbers is, it makes it a fraction." It is notable that he does not refer back to or draw inferences from the calculations he made in part (a), but instead he repeats them.

Finally, Adam also has difficulty expressing his general understanding of the phrase "M is divisible by N" in Question 3:

> Adam: OK, uh, M is divisible by N would, would for me, in my own words, I would say, it's it's like taking I, I, I would put it in numbers first, my own numbers I'd make up my own numbers to sh—, to kinda show, what I think, it would be easier. And I, I would like maybe choose like numbers like 9 and 3. And then I would say that 3, er 9 is divisible by 3 because uh, 3 can go in 9 equally it'll go 3 times 3 and it will equal 9.
>
> I: OK.
>
> Adam: So I guess, a way to say that, I would, uh, M is divisible by N because the w—, uh, N will go equally uh, amount of times into M.

Given his tendency to respond to problem situations with actions, it should not be surprising to see that he must again resort to actions in order to express his view of the divisibility relation.

In summary, we highlight two aspects of Adam's work. While the four parts of Question 1 concern the same multiplicative relationships expressed in different ways, Adam shows little awareness of this fact. He simply repeats many previous calculations in parts (c) and (d) rather than referring back to what he observed in parts (b) and (a), respectively. Rather than consciously exploiting the interrelatedness of the concepts, his approach is usually the "knowing through doing" characteristic of Stage I. He tends to view the issue as reporting the result of dividing (whether it "works" or not), rather than in terms of its purpose in determining the relationship between the numbers.

On the other hand, in dealing with each concept individually, he seems to be in transition from Stage I to Stage II; while he is disequilibrated by requests to reflect on his actions and explain or summarize their meaning in general, he does show a limited tendency to formulate predictions that are tested by immediate action. At this point in the interview, his grasp of the concepts appears not to be strong enough to allow him to respond with inferences rather than actions, so he has not reached Stage III in his response to divisibility tasks.

Understanding Divisibility and Prime Factorization

In this section, we consider the task of explaining why a number that is expressed in prime-factored form is divisible by any of its prime or composite divisors (as in Questions 4[a] and 4[b]). Being able to view a prime factorization as a product of a divisor with its quotient is one way to infer the answer to the question easily, without much computation. We propose that inferences of this type arise from reflecting on the results of coordinating the processes underlying the commutative and associative properties of multiplication[5] with one's process conception of prime factorization. If the individual associates divisibility primarily with division, making the inference may also include an application of the reverse relation between multiplication and division.

Determining Divisibility by the Action of "Looking"

We consider Alice's work on this situation in Question 4(a). In her responses, we learn that her apparent success is actually based on an observation of surface features, rather than on coordinating the necessary processes. Like some of the other students interviewed, Alice expresses the idea that one number is divisible by another number if the latter is "in there" (i.e., in the prime factorization of the first number); for her, to succeed on the task simply requires "looking."

Just prior to working on Question 4(a), Alice explained her definition of "M is divisible by N" in this way:

Alice: Um, well OK, then I'm just gonna say that I, that M being divisible by N um, you have to have a, you have to have an um, a whole, an equal number, I mean N has to go into M, oh wait, no, yeah, N has to go into M just a whole number of times.

I: Umm, hmm.

Alice: You know, you have to have just a whole number of times that it will go in there.

I: OK, yes, that's true. I noticed that when you did this you expressed it in terms of division, I think each time you say it in terms of division. Can you say anything about it in terms of multiplication?

Alice: Um, N times, I don't know, a certain whole number is equal to M.

However, when asked less than a minute later in Question 4(a) whether $M = 3^3 \times 5^2 \times 7$ is divisible by 7, she says the following:

Alice: Um, OK what I don't know what to do with this though. Well, I would say yes, because um what I always just look for is to see if um, the factors of this are 1 and 7, that means 1 times 7 equals 7.

I: Umm, hmm.

Alice: And they're both prime so you can't factor it out any more. And, um I just look in here, the $3^3 \times 5^2 \times 7$ to see if there is a 1 or a 7 in there.

I: Umm, hmm.

Alice: And since there is a 7 in there, then I say that yes it is divisible by 7.

I: OK, now how does that fit with what you said back here?

Alice: What I said about being divisible by?

I: Yes, because here you're telling me about looking in the number for something.

Alice: OK, well. [*pause*] OK, M [*pause*], OK, um, well if you have a s—, can I just work it out, and see if it is?

At that point, Alice changes M to decimal form, and then divides by 7 to check. Apparently, she has not made the coordination proposed above, since the only divisibility process that seems to be available to her here involves carrying out the indicated multiplication to return to decimal form.

Alice's stated approach was to "always just look" for the divisor in the given factorization of M. For someone whose actions are guided by an understanding of multiplicative structure, this action of "looking" at the prime factorization might be a meaningful, condensed version of the coordination proposed above. However, Alice is aware only that "looking" works, but not *why* it works—that is, her action is taken without an awareness of the mechanisms that guarantee its success. Once Alice has determined that M is divisible by 7, using division in decimal form, she comments:

Alice: See and it does go through. Well, let's find out why. [*laughs*]

I: [*laughs*] OK.

Alice: OK, well, see maybe this is just something that I don't know that much about because I, just that's the way to do it, is just that I factor it out and if those, if one of those factors is in there then it goes in . . . .

Notice that Alice distinguishes here between performing an action and knowing how or why the action produces success. We propose that she is now working at Stage II—she understands that there is more than knowing by doing, which allows the interplay of actions and conceptualization to begin. When she discusses the utility of the approach of changing M to decimal form, she shows her understanding of the advantages of making inferences over performing actions:

Alice: . . . but the thing is you would have to figure out why it does that because, in every case, I don't think you would do that, because um, this is small enough that you could multiply it out, but I mean, in many instances, you'll get it where it's so big that there's no way you can hold all those numbers in your calculator to actually divide it.

I: Right, right.

Alice: So you have to find like a shortcut, to figure out why it works. Um.

Disequilibrated by the need to apply her divisibility process to numbers in prime-factored form, she attempts to reconstruct it. Through a short discussion with the interviewer, she reaches the conclusion that M is divisible by 7 because of the related roles that the number 7 plays in the statements "M divided by 7 equals a whole number X," and "7 times X equals M." She acknowledges that this is a new understanding for her:

Alice: Yeah, well I just got it just now, 'cause I never understood like why you do it that way. I just know that it works that way.

It is worth pointing out that doing what Alice did initially—basing her reasoning on surface features instead of on an understanding of the properties of arithmetic operations—is not unusual among prospective elementary teachers. It appears elsewhere in our data, and often in our courses. Liping Ma (1999) documents the same kind of mathematical behavior in the responses of some experienced elementary school teachers as they considered a variety of topics in elementary mathematics. For example, the responses of some of the teachers interviewed by Ma revealed superficial views of the role of place value in a standard subtraction algorithm and the role of zero as a placeholder when finding partial products in a standard algorithm for multiplication.

Reconstructing a Divisibility Process

Next we consider Karyl's work on Questions 4(a) and 4(b), and argue that she is moving through Stage II to Stage III during this discussion. She has a strategy for carrying out the task, but when pressed for explanations, her responses reveal some weakness in her schema for multiplicative structure, with regard to the construction of divisibility processes for numbers in prime-factored form. We present her work in detail in hopes of providing some insight into how conceptual understanding of divisibility in this setting can develop.

Earlier in her interview, Karyl exhibited a comprehensive understanding of multiples, divisors, and divisibility for numbers in decimal form. She summarized, in writing, her understanding of what it means to say M is divisible by N. She formulated it first as M divided by N equals X, where the quotient X is a whole number, and then pointed out that, consequently, N times X equals M. She had used this formulation explicitly to guide her actions in a variety of examples in Questions 1, 2, and 3, but only for numbers expressed in decimal form.

Karyl worked confidently when the numbers were presented in decimal form; in contrast, her approach to working in prime-factored form was very tentative, and did not appear to be guided by the same depth of understanding, as illustrated in the excerpts below. Karyl is considering Question 4(a): Is $M = 3^3 \times 5^2 \times 7$ divisible by 7?

Karyl: I would say it is divisible by 7 because you can think of $3 \times 3 \times 3 \times 5 \times 5$ all that is multiplied you can think of kinda by 7 you could think about it that way and so this number would be divisible by 7 but I'm gonna check it on here 'cause when I do things that I'm not sure about I like to check it and make sure I'm not doing it incorrectly. OK, when I'm doing it on the calculator I did $3 \times 3 \times 3 \times 5 \times 5$ . . . times 7 and then that's gonna equal 4725 and then what I'm gonna do then is 4725 divided by 7 and that equals 675, so I just sometimes I do get confused with this kinda stuff too so I always like to check 'cause sometimes I think of the wrong things, so. [*laughs*]

In these and subsequent remarks about the factorization of M, Karyl refers to the factor complementary to 7 as "the $3^3 \times 5^2$," "all this," and "these whole numbers"—never as a single number. In fact, when asked about the quotient when M is divided by 7, she considers only converting M to decimal form and dividing by 7, obtaining 675. Referring to her earlier stated definition of divisibility of M by N, she notes: "I guess the only way I really did that, the division part, is when I checked it on the calculator."

To see whether she recognizes that the quotient upon division of M by 7 can be identified directly from the prime factorization, the interviewer continues:

I: And how is the 675 related to what you have up there? [the interviewer is referring to the given prime factorization of M]

Karyl: Well um, it's the 3^3 times 5^2 times 7 and then you, when you divide that by 7 that's the answer to that. . . . So I guess that's kinda just proof to say that it works. [*laughs*] 'Cause it comes out to a whole number.

Her response indicates that she is still thinking of converting to decimal form and dividing 4725 by 7 to obtain 675, rather than seeing the prime factorization of M as a product of 7 and 675. The interviewer continues with a more specific prompt:

> I: Umm, hmm. Do you know what the prime factorization of 675 is?
>
> Karyl: It would be the same as right here, it would be the 3, oh wait, no it wouldn't, let's see, you take, I think it would be the, let me see, $3 \times 3 \times 3$ times. . . . OK, it is it, OK 'cause you did divide by 7, I thought that's what it would be but I wanted to double check, so it's the $3^3 \times 5^2$ and that would be the 675, 'cause all you really did was kinda take away the 7.

Karyl has now identified the quotient upon division by 7 as part of the prime factorization—something that should have been obvious earlier if she were thinking about M as a product of 7 and the quotient $3^3 \times 5^2$. Her observation that you can "take away the 7" might indicate growing awareness of the effect of coordinating the reverse processes of multiplication and division by 7.

However, her ability to view the quotient and divisor as complementary factors of M is fleeting for, in the next question (which asks whether the same M is divisible by 5), she is unable to recall and apply this method again, without prompting. That is, her understanding of how the multiplicative structure of the natural numbers might apply to this task appears to be only for immediate application, and not as something that can direct her future actions, which we classify as Stage II behavior.

> Karyl: OK. Do you want, um, do you want me to do it without a calculator and just do it by looking at the numbers or do you want me to check my work or how should I do that?
>
> I: However you feel most comfortable.
>
> Karyl: Ok. I might check it at first if I'm not sure. OK. Just, I guess, um, let's see. [*mumbling, inaudible*] OK, so this is the same kinda question they're asking this time, it's just with different numbers and as the above, 'cause this is "is M divisible by 7" and now I'm doing "is M divisible by 5" and without using a calculator or anything I'd just look at this and have to say yes because of the um, 5's that uh, 5^2, er, when you, I guess when you look at the prime factorization since you see 5's in there I think about it as being divisible by 5 because it's *in there*, I guess [*both laugh*].

She goes on to say "I don't really know how to explain it," but she notes that it is unnecessary for her to actually do the division this time: she knows that M = 4725, and 4725 is clearly divisible by 5 since its ones digit is 5. With regard to explaining why M is divisible by 5 through analyzing the prime factorization, though, she still appears uneasy, saying:

> Karyl: I really, I don't know, I don't, I don't think it's that I don't know *how* to explain it, I mean, I think I know in my head what I'm thinking, but

I really don't know how to say it to make sense in mathematical sense, in mathematical terms [*laughs*].

The approach taken in the course Karyl has just taken is that making sense in "mathematical terms" includes making inferences from established principles; she may be acknowledging that her understanding of the task is not at that point yet. It appears that she understands the coordination that supports viewing M as a product of the divisor and quotient only through taking actions. Her confusion here suggests that she is not confident or aware enough of the mechanism to recall and apply it again, without additional direction.

The interviewer directs Karyl back to her own definition of divisibility given in her response to Question 3:

I: Well, again, back here you said what it meant for M to be divisible by N.

Karyl: Oh, maybe that might help me if I look at that again 'cause like, OK . . . M divisible by N, OK so it's, you can think of, the 4725, is that divisible by 5 . . . [*divides 4725 by 5, using a calculator*] . . . and then I guess, because you know that it *is* true because it's a whole number times 5 equals the dividend . . . so it's the same thing kind of again so I guess you can think about it in that way.

The interviewer continues by asking her what the quotient would be in this case "without using a calculator." She still does not see this as a question that can be answered from the prime factorization directly, but instead proceeds to divide 4725 by 5 (using paper and pencil, long division). As before, she may not be seeing the factors other than 5 as forming a single number. In terms of mental constructions, we propose that she is having some difficulty coordinating the processes underlying commutativity and associativity with the process underlying the prime factorization. The interviewer then asks a more precise question about finding the quotient:

I: Could you do it by looking at the prime factorization?

Karyl: Oh I guess that you could really. I think. Let me double check this again, I think you take the 3 times 3 times 3, I think, times one 5 and times 7 [*inaudible*] that would be the 945, so you, I guess you can basically think about it as the 3 times 3 times 3 . . . all you gotta do is really take out, all you really need to do is take out one 5 out of the prime factorization because you're dividing by 5 so, you're not any, you're no longer multiplying because you're dividing.

For her, the key to seeing M as a product of the divisor and quotient seems to be captured in her statement "You're no longer multiplying because you're dividing." In terms of mental constructions, her observation could be interpreted this way: now that she sees the quotient X as a single number, she can de-encapsulate the prime factorization of M to the process of multiplying X by 5. She coordinates this process with division of M by 5 and recognizes that X is the quotient because division by 5 and multiplication by 5 are reverse processes (and so cancel out). That is, instead of carrying out (X × 5) ÷ 5, she reconstructs it to the process X × (5 ÷ 5).

This interpretation of her actions is supported by her focus on identifying the quotient in the prime factorization in subsequent responses. Moving now into Stage III on this task, Karyl has consciously formulated a meaningful process for divisibility questions for numbers in prime-factored form, and she is able to use it to guide her actions and inferences concerning divisibility of M by 9, and by 63.

> Karyl: And I believe the 9 will work because you can, as I said before, you took a 5 out of here, and this way you can take I think two 3s out of here and it will work and you can think about it as 3 × 5 × 5 × 7 . . .

The phrase "you can think about it as 3 × 5 × 5 × 7" might indicate that she is now thinking about M in terms of the combined effect of multiplying and dividing the number by 9. Interestingly, there is some regression to Stage II as she continues, for she persists in verifying her work by performing division in decimal form (for divisibility by 9 and by 63). The difference now is that the division is performed after she has identified the quotient in prime-factored form. Only when she considers the final question of whether M is divisible by 15 is she sure enough of her mode of reasoning to say:

> Karyl: And next is 15 and I probably won't even need to check this because I think I'm confident enough now that I know what I'm doing that I don't need to use that. . . . So this is a 3 and a 5, and then well you can break 15 down into 3 and 5 and then you can look at that, and the prime factors 3 and 5 are common in the dividend, there's 3 and a 5 in there so you can think of them as 3 × 3 × 5 × 7 so, 3 × 3 × 5 × 7 that equals 315, so the answer would be 315, and so that's gonna work.

Even here she focuses on the "answer"—the quotient is the number that you multiply and divide by 15. This may reflect how important recognizing the quotient in the prime factorization is, in her understanding of how her knowledge of multiplicative structure can be applied to answer questions of divisibility.

Karyl has moved from success to understanding with this task. She appears to have conceptualized the task in terms of coordinating multiplication, division, and prime factorization, facilitating her application of her understanding of multiplicative structure as a conscious guide to her reasoning and actions. No longer is her conceptualization of divisibility for numbers in prime-factored form on the periphery, as something she considers in retrospect after performing arithmetic actions in decimal form. Her work illustrates one way that learners might move from action-based decisions about divisibility to inferences based on an understanding of multiplicative structure.

Multiples and Prime Factorization

We consider in this section how the lack of awareness of a role for multiplication in generating multiples can limit an individual's ability to coordinate her or his schemas for multiplication and prime factorization to construct an understanding of multiplc in primc-factored form.

Questions 4(c) and 4(d) require the ability to consider multiples and non-multiples in prime-factored form. With some effort and occasional strong prompting by the interviewer, nine of the ten students we interviewed were able to solve these problems. Kathy was the only exception. To start with, we note that Kathy's view of multiple is the following (excerpted from her response to Question 2):

> Kathy: I would say a multiple is just repeated addition of that number . . . of just repeating that number over and over and over again.

This response indicates that she has a process conception of multiple, in terms of repeated addition, not multiplication. In fact, a review of her performance throughout the interview reveals that she strongly associates the concept of multiple with a process that continues indefinitely. Her conception is consistent in terms of repeated addition, which she sometimes refers to inaccurately as "doubling." For example, she says "I just remember that multiples are like 2, 4, 6, 8, or 3, 6, 9, like, it's just, you're multiplying in your head, you just, you know, doubling the numbers." Though she uses the word "multiplying" here, there is no evidence, at any point in the interview that she understands multiplication as the process of keeping track of the number of repetitions of an addend.

Moreover, Kathy frequently discusses multiples in ways that do not refer to multiplication or division, such as reasoning (in her response to Question 1[b]) that 2400 is a multiple of 24 because "you just take away the zeroes." When the interviewer asks her what must be multiplied by 24 to get 2400, she is unable to answer, even though both she and the interviewer read 2400 as "twenty-four hundred." When she begins to use her calculator, there is this exchange:

> I: What were you trying?
>
> Kathy: I was trying, I was yeah, I was just, that wasn't right though, but I just knew it immediately when I went 24 times 24 'cause that isn't what a multiple is, it's, you're doubling it, like 24 and 24, it's like repeated addition.

She reiterates that she understands the process to require adding 24 repeatedly, and that she knows "you would get to 2400 just because 24 is in there and you just take away the zeros." The interviewer tried several times subsequently to prompt her to consider multiples in terms of multiplication but she repeatedly deflected those suggestions.

We consider now Kathy's work on Question 4(c): Is $3^2 \times 5 \times 7^3$ a multiple of $M = 3^3 \times 5^2 \times 7$? In her work, initially we see a focus on the differences in the exponents on the primes, without any reference to viewing the relationship between the two numbers in terms of arithmetic operations:

Kathy: OK. [*pause*] I don't know why I'm thinking this in my head, but this is the first thing that popped in my head and I don't, I really don't think it has anything to do with what I'm thinking about, but um . . .

I: Tell me though.

Kathy: I just thought of it like, OK, there's one difference in the 3s . . .

I: OK.

Kathy: . . . there's one difference in the 5, and there's 2 differences or 3 differences, 'cause it's only, or 2 . . . 'cause this is this 7^3 and this is just 7^1 so there's 2 differences . . .

I: OK.

Kathy: . . . so then I, I thought of the differences but, um, that's really not gonna help to find the multiple [*pause*].

When the interviewer refers Kathy back to their discussion in Question 1, reminding her that they considered that each multiple of 24 as "24 times some whole number," she considers a possible role of multiplication and concludes, incorrectly, that the given number is a multiple of M:

Kathy: All right, so this would be yes, just because you're just multiplying one more to each so, 'cause I was thinking like, um, you just added, or you just put like another 3 on there . . . and since you're multiplying that one would be OK . . . and then 5 times 5 would just be another 'cause that's like another multiple.

I: Uh, huh.

Kathy: 5 times 5, and then 7 times 7 times 7, it would, so it would be a multiple because you're just, you'd have to make sure that they're not adding, but that it, this is both multiplication and you're just doubling it like you did with this, like 24, 48, you know 96, 'cause you're just you know going 3 times 3 . . . you know 5 times 5, so yeah, I would say that yes, it is a multiple.

Kathy's confusion about the role of multiplication in this task can be seen in her mixture of references to addition and multiplication, and in her failure to distinguish the respective roles of the two numbers. Her comment "You'd have to make sure that they're not adding" actually might not refer to the operation of addition. Rather, we think it might refer to her belief that primes which are not factors of M, cannot be factors of its multiples (that is, there are no *additional* factors), which is evident in her response to Question 4(d): Is $3^4 \times 5^5 \times 7^3 \times 13^{18}$ a multiple of $M = 3^3 \times 5^2 \times 7$?

Kathy: I'm not really sure because I don't know if that, I don't think that makes sense because even though they're all multiples and you're keep, you keep going up . . . um [*pause*] I, I'm not sure if that, that would be right, just because uh, I guess, this, it was easier to see with this one [referring to Question 4(c)].

I: Umm, hmm.

Kathy: The 13 throws me off just because, I'm not sure if um, 'cause see, I don't think it would be, just because the 3, the 5, the 7, this kept going on, it just kept going up and it fit in with my pattern of, you know a multiple, yeah, it can have an infinite amount as, you could just keep going and going and going . . . with the same numbers, but with that 13 in there, the 13 keeps going and going but it wasn't from the original.

She goes on to conclude, incorrectly, that the given number is not a multiple of M. Her comments indicate that she has in mind some process or "pattern" for generating multiples in prime-factored form: it is a process that continues indefinitely "with the same numbers." It is worth noting that this process does produce correct decisions in a limited selection of cases, but not in general.

We also propose that Kathy's process does not involve an application of either addition or multiplication to the prime factorizations. For, if she were thinking about repeated addition, she might refer to repeatedly adding the prime factorization of M, and ask how multiples could be recognized. If she were thinking about whether the number could be obtained by multiplying M by another natural number, we would not expect the factor of 13^{18} to be an issue for her. Since Kathy appears unable to apply operations to prime factorizations in a meaningful way, it could be that she does not have an object conception of prime factorization. Indeed, earlier in her interview, she admits that she prefers to revert to decimal form, where she can "see the numbers." We interpret this view of prime factorization as indicating a process conception, since the factorization is perceived only as a string of instructions for producing a number in ordinary decimal form.

To summarize, Kathy's responses to 4(c) and 4(d) fall squarely in Stage I. Much of her reasoning is based on her observations of surface features of mathematical notation such as differences in exponents and the appearance or nonappearance of particular numerals. Her conceptions may have been formed by previous successes, since such observations will give correct answers in limited contexts. The failure to advance to a more useful conception of multiple appears to be caused, at least in part, by her lack of awareness of the role of multiplication in generating multiples. Without that understanding, her reasoning is focused on what she sees, instead of on the general mechanisms that explain how what she sees was produced.

Constructing the Least Common Multiple

The concept of the least common multiple of a pair of numbers is constructed through applying one's schema for multiplicative structure to the problem of finding the smallest natural number that is a multiple of each of a pair of numbers. An application of one's schema could result in the construction of a variety of procedures. In our interviews, we observed subjects using the following three approaches:

Set intersection: In this method, one creates an ordered list of consecutive multiples for each number and finds the first one that appears in both lists. Since the lists of multiples are generated in an ordered fashion, the fact that the least common multiple has been found is obvious.

Create a multiple and divide: In this method, one creates an ordered list of consecutive multiples of one number while simultaneously checking each new entry for divisibility by the second number. For one who has constructed the relationship between multiples of N and divisibility by N, inferring that the least common multiple is produced is immediate. It is significant that this method was used by participants only in cases where the numbers were expressed in ordinary decimal form.

Prime-factorization: This is an algorithm that typically appears in the textbooks for this course, and it requires one to compare pair-wise the prime powers in the prime factorizations of A and B, and construct a prime factorization of their least common multiple. Seeing that the procedure actually produces the least common multiple of A and B is problematic for many students (we discuss this issue in the next section).

Adam's work will show his being able to compute a least common multiple in various settings and being able to recognize the concept in all of its forms is not necessarily equivalent. We used three different tasks that involved the concept of least common multiple: Question 5 used the paraphrase "smallest counting number that is a multiple of each," Question 6 used a physical context that could be modeled by the least common multiple, while only Question 7 used the wording "least common multiple," which was familiar to the students.

We believe that Adam's work on Questions 5 and 6 indicates a conscious application of his knowledge of multiplicative structure, while his application of the prime-factorization method for the LCM in Question 7 is another example of his action-oriented applications. In starting Question 5, he says:

Adam: . . . a way to do it, since I can't think of one right off the bat would be, and it would be a long time probably for me to find the answer, would be like, I could just start multiplying 378 times 2 and see what that answer is and then divide that by 72 to see if it goes evenly. I could do it that way . . . Uh . . .

He easily solves the problem by generating successive multiples of 378, dividing each time by 72, effectively coordinating a process of generating multiples through multiplication with his conception of multiple in terms of division. Nonetheless, he comments at the end of his solution:

Adam: . . . but I know there's an easier way, or a different way to do it.

When he turns to Question 6, he recognizes the context as structurally similar to the situation in Question 5, and says:

Adam: . . . Uh, for this one it's kinda like [the previous question], where you have to find the smallest counting, counting multiple, um, for 168 and 108 that's what you're doing each time, each thing's going up by that much.

Adam used the same method as in the previous problem. It was an easy application and he clearly explained his reasoning (noting that 72 and 378 are divisors of the number he found, 1512). He never used the terminology "LCM" or "least common multiple" in his discussion, and instead paraphrased the statement of Question 5 to say that he found the "smallest counting multiple" [sic]. The interviewer asked him directly, at the end of his response to Question 6, whether he recalled another way to solve these problems, and he stated twice that he did not. In both problems, we see a conscious application of his knowledge of multiplicative structure guiding his actions. He states and follows a plan, so his actions are not *ad hoc*, and he understands and can justify the steps.

In the case of the reversal required in Question 7, Adam's inability to apply a direct action proves to be an obstacle, as might be expected for an individual who prefers action as his first response:

Adam: OK. [*reading question*] Well, I, I, I could I um, I know I'm not gonna find the right answer, I know I won't if I wanted to just start plugging in numbers back and forth and see, see what will work, um but I don't remember how to do it at all.

His reluctance to consider a process of "trial and error" suggests that he does not have any criterion in mind that would suggest possible values for x and y to yield $\text{LCM}(x,y) = 200$. In particular, although he has shown in his work on Questions 5 and 6 that he understands that a common multiple of x and y has x and y as divisors, he does not recall and apply this condition here. After a few minutes of little progress, the interviewer asks him how he would find the LCM of two numbers, if the numbers were given. He responds:

Adam: Uh, well the way the way I always find the LCM is by, I prime factor them all the way down.

He proceeds to demonstrate the prime-factorization method for finding the least common multiple as the method he "always" applies. Of course, his statement is directly contradicted by his actions on the previous two tasks. One way to interpret this paradox is that, for him, the "LCM" is a specific object that is constructed by encapsulating a particular method. The phrase "LCM," which was first uttered by the interviewer, triggered de-encapsulation to this prime-factorization method, but the concept of the "smallest counting number that is a multiple of both" did not.

Further evidence that Adam views the LCM and the "smallest counting number that is a multiple of both" as separate objects can be observed in his work in Question 7, where he repeatedly used the prime-factorization approach, even to find LCM(50, 100). Here, he first writes $50 = 2 \times 5^2$ and $100 = 2^2 \times 5^2$:

I: OK, so Adam chose 100 and 50 and he's checking whether or not they are two numbers that have an LCM of 200. [The interviewer has just turned over the audiotape and makes this comment for documentation purposes.]

Adam: [*long pause, working*] Yeah, that's what I found is working, they, I mean they do have an LCM of, I think, yeah, 'cause you take from the way I do it I, you do 2^2 and 5^2 and I take the bigger of the two, and 2^2 is there and since those are equal I'll just take one of them, which is 5^2, 'cause—

I: So what's 2^2 times 5^2, is it 200? 2^2 times 5^2 looks like, looks like 100.

Adam: Oh it's 100, that's right.

It is not unreasonable to suggest that if he had been thinking of the "smallest counting number that is a multiple of both 50 and 100," he would have immediately seen that the result would be 100, and no application of this procedure would have been necessary.

To summarize, Adam can consciously apply his knowledge of multiplicative structure to construct a concept of the least common multiple—this is precisely what he brought to bear on Questions 5 and 6. However, he apparently has a separate concept of the object produced by the prime factorization method. This method is not constructed through a conscious application of his understanding of multiplicative structure, but rather is a memorized procedure.

Apparently, students will not necessarily develop a broadly applicable, unified conception of least common multiple merely by working through a variety of examples. Reflection on actions and processes is required in order to develop such a unified conception. In this case, Adam solves all three problems with only mild prompting, but his work strongly suggests that he does not realize that the three situations relate to the same concept.

Success with and Understanding of the Prime-Factorization Method for LCMs

The prime-factorization method for finding least common multiples is a relatively easy algorithm to apply. However, as teachers, we have noticed that few of our students can explain why this procedure actually produces the least common multiple, even when they are successful in applying it. In part, it was this observation that inspired us to study prospective elementary teachers' conceptions of divisibility.

In mathematics courses for elementary teachers, the unique factorization of natural numbers guaranteed by the Fundamental Theorem of Arithmetic (FTA) is typically accepted without proof (see Zazkis & Campbell, 1996b). In our courses, we expected arguments concerning divisibility and prime factorization to take the theorem as a starting point. Thus, in our view an "understanding" of the prime-factorization method for finding a least common multiple could be based on acceptance of the FTA, without insisting that the student could prove the theorem. This is the sense in which we suggest that a schema for unique factorization of natural numbers into primes can be part of a college student's schema for multiplicative structure. At the same time, we acknowledge that it is not a safe assumption that college students actually understand the most basic implications of the theorem (see results in Zazkis & Campbell, 1996b).

To justify the prime-factorization method for finding LCM(A,B), what we might reasonably expect from our students is the following, stated informally: the prime factorization of a natural number is part of the prime factorization of any of its positive multiples. Therefore, the prime factorization of any common multiple of natural numbers A and B must contain the prime factorizations of both A and B. Now, LCM(A,B) is the smallest natural number with that property, in the sense that dividing LCM(A,B) by any of its prime factors produces a natural number that is not a common multiple of A and B. The "choose the higher exponent" rule that typically appears in textbooks for this course is a result of this reasoning.

Next we consider the mental constructions that might support the ability to construct such an explanation. We hypothesize that it consists in first coordinating the concept of multiplication with the concept of prime factorization to construct the process of generating a multiple in prime-factored form. This might be considered to be a reconstruction of one's process of generating multiples to accommodate generating multiples in prime-factored form. Through reflecting on this process, the individual may infer the condition that a prime factorization of a multiple of a natural number N must contain the prime factorization of N. To construct the concept of common multiple in prime-factored form requires comparing the prime factorizations of (imagined) multiples of each number. Because comparison is an action, the multiples process must be encapsulated as an object. Through reflecting on this comparison, the individual may infer the condition that the prime factorization of a common multiple must contain the prime factorizations of both numbers. Inferring the properties that the least common multiple has might involve reflecting on the action of dividing a common multiple by any of the primes in its factorization (or, more simply, just removing one of the primes) and deciding when the quotient would fail to be a common multiple.

We present the responses of four students attempting to explain the prime-factorization method for finding LCMs. The first two students are responding to Question 5, where they are finding LCM(72, 378) using the prime factorizations $72 = 2^3 \times 3^2$ and $378 = 2 \times 3^3 \times 7$.

Manipulating Surface Features

Kathy has applied the algorithm to determine that the least common multiple is $2^3 \times 3^3 \times 7$, and then the interviewer asks her what she was thinking about as she obtained it. She notes that she started with the prime factorization "to find what factors go into it," and then explains how she chooses the "greatest amount" when she compares prime powers of the two numbers. As to why she chooses the greater exponent in each case, she shows confusion and then admits:

> Kathy: I don't know how to explain it, because with the multiples you just, you just *have* to! [*laughs*]

When the interviewer later asks her how she knows that $2^3 \times 3^3 \times 7$ is a multiple of each of the numbers 72 and 378, she states that the main reason is that "the factors are the same." When the interviewer points out that 7 is a factor of 378, but not 72, and asks her why a factor of 7 shows up in the least common multiple, her final analysis is as follows:

> Kathy: [*pause*] I would say maybe because um, the 7 would be in there because the 378 is a greater number and since it's a greater number it needs more um, factors in it, um, and multiplication just doubles and doubles so, [*inaudible*] um, I'm not sure, we can come back, I guess. I'll think about that.

It is not surprising that Kathy has difficulty with the concept of common multiple in prime-factored form, since we saw earlier that she has inadequately constructed the concept of multiple in prime-factored form. She has a Stage I success here: applying the algorithm correctly requires only a manipulation of surface features, not an understanding of its rationale.

Using Trial and Error

We now turn to Alice's work on the same question. In Alice's response to Questions 4(c) and 4(d), earlier in the interview, she discussed the role of multiplication in recognizing when one number in prime-factored form is a multiple of another. While she doesn't have an algorithm in mind for this task, her process conception of multiple supports a trial and error approach to the task that allows her to consider whether each of her several proposed solutions could be a common multiple of the numbers.

On her first attempt at Question 5, Alice proposes an answer of $7 \times 2^4 \times 3^2 \times 7$. She describes an approach of writing down and crossing off prime factors that has no connection to the concept of multiple that we could identify. When asked for clarification, she acknowledges that she doesn't have a clearly defined procedure in mind:

> Alice: Oh I don't know a rule, I'm just. . . . Um, I don't know, I wasn't really, I know we did a couple of problems like this on the homework and I, I don't think I was really that sure how I was doing it.

The interviewer tries to refocus her attention on the task, by emphasizing the need to construct a common multiple:

> I: Well, let's look at the question again, let's see. "Smallest counting number that is a multiple of both 72 and 378." So you need to build a multiple of those 2 numbers.
>
> Alice: Like um, $2^4 \times 3^5 \times 7$? Is that what you mean?
>
> I: Is that a multiple of both?
>
> Alice: Well.
>
> I: How did you get that number by the way?
>
> Alice: This number right here? [*she points at* $2^4 \times 3^5 \times 7$]
>
> I: Yes.
>
> Alice: I just multiplied um, my prime factors together.
>
> I: Uh, huh. So that would certainly make a multiple.
>
> Alice: Yeah, but that might be too big.

Alice understands how to find at least one common multiple, and that not just any multiple will do. As she continues, though, it becomes clear that she is not structuring her actions with a clear concept of common multiple as a guide. While she is able to consider whether one given number is a multiple of another, the coordination required to produce a common multiple (in general) eludes her. She devises another plan:

> Alice: Oh, OK here, I'm getting it now. This is what I think I'm gonna do. I'm gonna take my 2 prime factors, 2 sets of prime factorization together and I'm gonna get, I'm gonna put $2^2 \times 3 \times 7$, do you see what I did? I took this 2 and subtracted it from 2^3 and I so got 2^2 then I took this from this and I just got 3. And then I put my 7 down.
>
> I: OK, now is that number a multiple of those two numbers? [*The interviewer is referring to her answer of* $2^2 \times 3 \times 7$.]

Alice: No.

I: How come?

Alice: Because . . . [*pause*] no, it's not, because you have to multiply this times something to get this and you can't do that.

I: Hmm.

Alice: Do you understand what I'm saying?

I: Yes. Umm, hmm.

Alice: Um. It's not a multiple of this either because you'd have to multiply, multiply this times something to get this and you can't do that either.

I: Yes.

Alice: OK, I'm confused. [*laughs*]

Her comments here indicate that she is working with a process conception of multiple, since she is focusing on the process of multiplying, rather than on the structure of a multiple. She may be "confused" here because she is trying to apply the action of comparison to processes of generating multiples, rather than to objects that are multiples.

In a few seconds, she proposes another possibility, $2^2 \times 3^2 \times 7$, and the interviewer asks her whether it is a multiple of both numbers. Again, she sees immediately that it is not. To answer the interviewer's question about how she knows it is not a common multiple, she makes an adjustment based on an inference arising from an action of comparison of the prime factorizations of 72 and 378 (rather than of their imagined multiples):

Alice: Because I had two to the square, I had 2^2 and then I know that, what am I gonna multiply 2^3 by to get 2^2.

I: OK, so having 2^2 there is not enough.

Alice: No, so I just changed it to 2^4.

I: Why did you change it to 2^4?

Alice: Because, um, I have 2^3 over here.

I: Uhh, huh.

Alice: And then I have a 2 over here so I just thought I would add them together. I mean I would put this 2 in with this 2. Do you understand what I'm saying?

By "add them together," she may be referring to the exponents (on 2^3 in the factorization of 72, and on 2^1 in the factorization of 378). The interviewer encourages her to pursue the strategy of comparison:

I: Well, if you just focus on this one for a minute. [*The interviewer points to the prime factorization of 72.*]

Alice: Umm, hmm.

I: That you know that, for this one, you have to have 2 to what power?

Alice: 2^3.

I: 2^3. Will that be enough for this one? [*The interviewer points to the prime factorization of 378.*]

Alice: What do you mean, what do you mean?

While the interviewer's hint is vague, the confusion Alice expresses might indicate that she is not aware of the purpose of the comparison in this context. After a few additional exchanges in which Alice confuses the direction of the multiplication and is further prompted by the interviewer to relate the comparison of exponents on the powers of 2 to multiplication, she exclaims:

Alice: And then, oh, I think I'm getting it. OK, um, 3^3 because, oooh that'll work, that'll work, because look, OK, if you take 7, if you take this, this, this will work because I can multiply this by 2^2, and that will give me this, and then this by nothing, so if I multiply this times the 2^2 this will give me this, if I multiply this times 7 times 3 this will give me this.

At the time of this last comment, she wrote $2^3 \times 3^3 \times 7$, and indicated that she could obtain it by multiplying the prime factorization of 378 by 2^2, and also by multiplying the prime factorization of 72 by 7 times 3.

Looking back at her work, Alice comments on the trial-and-error approach that she used, both here and in her homework.

Alice: Well, I think it was just like I wasn't really sure in the first place how I did it on my homework I just kind of like, worked around until I got the right answer.

I: Uh, huh.

Alice: You know.

I: Oh you didn't have a particular rule you were following.

Alice: No. Uhh, huh.

I: So, do you think, can you say what a rule is, at this point?

Alice: You need to find, well, just kind of like it says in the thing, just find an answer that is a multiple, you know the multiple, like take these times something, and then this is a, this is a multiple of this and it has to be the smallest one.

Alice's last comment might indicate how she finally views her response to the task. Rather than describing a specific procedure, she responds to the suggestion of giving a "rule" by making only a general reference to what she was trying to find. We interpret this as an example of early Stage II work, where action is, at times, ahead of conceptualization. Piaget (1978, p. 120) discusses situations in which "momentary" discoveries give rise to very unstable memories of precisely how the success was achieved, particularly when the discovery is facilitated by suggestions. Alice's action of comparison took place in a haphazard way, and the prompts of the interviewer played an important role. Alice can neither describe the aspects of the comparison specifically, nor does she use this idea later in the interview when she has the opportunity. Moreover, because each application of her schema for multiplicative structure resulted in immediate action, rather than in reflection or reasoning, the procedure she constructed to complete the task probably remains, at best, an action for her.

Making Inferences about an Algorithm

Next, we turn to Ed's work on Question 6. He has just identified the least common multiple of 168 and 108 to be $2^3 \times 3^3 \times 7$, using the prime factorizations $168 = 2^3 \times 3 \times 7$ and $108 = 2^2 \times 3^3$.

I: I think I heard you say the rule is: you take the largest of each [exponent], is that what you said?

Ed: Yeah basically, uh, the largest, we have a choice between 2^2 and 2^2, so you want to take 2^3, so you have a choice between 3 and the 3^3 so you take 3^3 and then, then 7 is, is over here in 168 and it doesn't even exist in 108 so you put 7 in there.

I: Um, hmm. Why do you need the largest one?

Ed: Well, because if you don't have, if you don't have, all, all the possible factors in there, you're not going to get larger, the large enough number to come out with your answer.

At this point, Ed appears to be referring to the size of the number the procedure will yield, rather than its structure as a common multiple. The interviewer probes on the issue of the sense in which it is "large enough." He responds:

Ed: If it's smaller than that, it's not going to be large enough to be the least common multiple, it's not going to be large enough that, I mean you don't want it too large I mean, obviously, or it won't be the *least* common multiple.

I: It wouldn't—

Ed: It might be the second least common multiple or something like that.

I: Or the second common multiple.

Ed: Yeah, or the third or fourth or, or just nothing. Or it might not be that at all, but if you, if you don't have the largest, then it's going to be just the opposite problem, it'll be too small.

We see two things in this exchange. One, by suggesting he could get "just nothing," it appears that he doesn't realize that choosing the exponents as he's describing will always produce a common multiple. And second, he is still focusing on the size of the resulting number without relating it to the property of being a common multiple or not. The interviewer provides a prompt to focus on the size of the exponents instead of the size of the resulting number:

I: Ok, so you're looking for something that's a multiple of both of those?

Ed: Yeah, basically.

I: And so that way of choosing the exponents you have is to make sure that the exponents are big enough?

Ed: Right, right, yeah, these have, these have both got to be factors of the problem. [*In his written work, he circled the factorizations of 168 and 108.*]

Ed doesn't directly address the issue of why $2^3 \times 3^3 \times 7$ is a common multiple of 168 and 108, although he hints at it by pointing out that the factors of the two numbers must appear in its prime factorization. While he is somewhat inarticulate in expressing his thoughts, he is reasoning about why the task is done the way that it is. Justifying his choices through reasoning instead of through computing indicates, in our view, that he is beginning to work at Stage III; that is, his reasoning is consciously guided, if incompletely, by aspects of his understanding of multiplicative structure. However, in his later work in Question 7, there was one point at which he considered it possible for LCM(x,y) to equal 200 without both x and y being factors of 200, so his awareness of the necessity of this property appears to be unstable.

Reasoning about Multiplicative Structure

Finally, Patti is a student who can apply her knowledge of multiplicative structure consciously and deliberately in explaining how and why the prime factorization method for LCM works. Like Ed, she identifies the least common multiple of 168 and 108 to be $2^3 \times 3^3 \times 7$, using the prime factorizations $168 = 2^3 \times 3 \times 7$ and $108 = 2^2 \times 3^3$.

Patti describes how she chooses the higher exponent on common prime factors, and then automatically includes the prime power for any prime that is not common. She never directly explains the latter choice, although this issue is covered by her later inferences. Her comments contain several examples of inferential reasoning that draw consciously on the coherent structure of her schema for multiplicative structure. She begins by verifying that she has found a common multiple:

Patti: Well, I know it's a multiple of 108 because, um, 2^1 uh, times 7 would give me this LCM, OK, which is, I mean 2^1 times 7 times the prime factorization of 108 would give me this LCM.

I: OK.

Patti: And then on 168, um, 3 to the, 3^2 times, the prime factorization of 168 would give me this LCM.

In this response to the interviewer's question about how she knows she has found the *least* common multiple, Patti relates the prime factorizations of the two numbers to the prime factorizations of the LCM, relates divisibility to multiples, and comments on how her actions guarantee minimality:

> Patti: Because, um, it has to, it has to have these numbers in there, and it, it's the least because I didn't jump up to like big [illegible] could divide into, I took the least that they both can divide into, it's just the bigger of the two.

The strength of the process she has constructed for this task is then tested in Question 7, which requires a reversal of one's process for finding the LCM. She starts by finding the prime factorization of 200, and then after a pause, chooses 2^3 and 5^2 as the first pair. Asked for an explanation, she points out that because the two numbers have no common factors, their product is the LCM, since both the 2^3 and the 5^2 must be chosen. The interviewer then asks her to find another pair:

> Patti: OK, um, let's see [*pause*] OK, um 2^3 times 5^1 would be one number, and 2^2 times 5^2 would be the other, because as I said before, you see what's in common, and both the 2s and the 5s are, are you know, in common, they both have that, but you go for the higher exponent, and so you would go for 2^3 in the first number and multiply times 5^2 in the second number, and you would come up with the same factorization as 200.

A further indication of Patti's understanding is her ability to discuss necessary, as well as sufficient, conditions on the algorithm. Below, she offers her reasoning that no factors other than 2 and 5 need be considered, basing her explanation on an inferred result of the procedure in the case that such factors were introduced:

> Patti: . . . if you introduce any other numbers, you have, see like if I had 2^3 times 5^2 times 7 say, and down here I had 2^2 times 5^2 times 3^1 say, . . . then 7 and 3 have to be dealt with, and in order for them to be dealt with they have to be multiplied times, say 2^3 and 5^2, and once you've done that you've muddied it all up and it's not 200 anymore.

Asked for yet another pair, she chooses 2^1 times 5^2 paired with 2^3 times 5^1, and then, without being asked, volunteers a description of how her method ensures that the two numbers are smaller than 200:

> Patti: So that's why you have to . . . you have to juggle 'em around a little bit, so that you have, you have to have the 2 with the higher exponent in one of the numbers and the 5 with the higher exponent in the other number or

> vice versa, you know, which ever way, but you can't have . . . both of them together because then you'll come up with um, 200.

We interpret Patti's work to be at Stage III since she is aware of, and explicit about, the conditions she considers in constructing a process to carry out the task. More significantly, her discussion is based on inferential reasoning that concentrates on the reasons that certain actions are guaranteed to succeed and why other actions cannot.

DISCUSSION

The general goal of this study was to learn more about how prospective elementary teachers apply their schemas for multiplicative structure to divisibility tasks. Our intention is to use this knowledge in future efforts to formulate a more complete and useful analysis of the divisibility schema. The task of elucidating the structure of the divisibility schema requires understanding both how the fundamental concepts are constructed and how they might be related to each other in forming the underlying coherent structure of a schema. One contribution of this study is the formulation of preliminary APOS analyses of certain number theoretic concepts that were presented in the cases. In addition, while the issue of how the divisibility schema develops is still unresolved, we believe that our use of the stage model helps move us forward, as will be explained.

In APOS theory, a collection of actions, processes, objects, and other schemas is referred to as a schema only if the collection is linked in a *coherent* framework. Consequently, a central theoretical issue in analyzing a particular schema is determining what constitutes coherence. In general, coherence has to do with being able to determine what is within the scope of the schema, knowing what the capabilities of the schema are, and being able to recognize the situations in which the schema can be applied.

As proposed earlier, the components of an individual's schema for multiplicative structure might include schemas for the arithmetic operations, schemas for their properties, and a schema for the unique factorization of natural numbers as the product of primes. We claim that applying one's schema for multiplicative structure to divisibility tasks may lead one, over time, to constructing a schema for divisibility; that is, our view is that the divisibility schema has its genesis in one's schema for multiplicative structure. The stage model is useful in that it provides a way to begin to analyze how, and to what extent, an individual explains the concepts of divisibility and the links among them by consciously applying her or his understanding of multiplicative structure. We conjecture that an awareness of the roles played by the various aspects of multiplicative structure is an essential first step in developing the coherence necessary to have a schema for divisibility.

For example, consider the responses of Kathy and Alice that were presented earlier. Both tend to respond initially to the tasks at Stage I, in that the first explanations they offer are based on surface features. But Kathy does not appear to move from Stage I, while Alice does. As we review their performance during the interview, one difference is clear: when in conflict, Alice looks to multiplication and its properties for an explanation, while Kathy shows little awareness of the possible roles of operations, even when prompted. While Alice's efforts are uneven, she seems to grasp that the explanations lie in an understanding of operations and their properties. There is no evidence that Kathy has this kind of insight. On the basis of this data, we would conjecture that Alice has made more progress toward developing a divisibility schema than Kathy has.

Adam's tendency to respond to problem situations with actions provides a second issue to consider. His strong preference for procedures seems to correlate with a tendency not to reflect on actions and interiorize them as processes. This had noticeable effects on his work early in the interview: first, the degree of success he has with tasks through "knowing by doing" obviated the need for reflection on his actions concerning divisors and multiples (until he was forced to, in order to respond to Question 2); second, it leads to some dependence on the use of a calculator. Since division is easily carried out on a calculator without either the time or the need for reflection, the importance of thinking of division as the reverse of multiplication is diminished. Adam's concentration on division, coupled with a lack of awareness of the role of multiplication, suggests that development of the coherence necessary for a divisibility schema might be problematic for him.

Finally, there were points at which the analysis of the data was clarified by our ability to coordinate an APOS analysis with the criteria expressed in the stage model. In some situations, we concluded that the ability to pass from Stage II to Stage III depended on the individual being able to view a particular concept as a cognitive object (which is reasonable, since the ability to draw inferences requires one's cognizance of the relationships between objects). For instance, this was the crucial difference between Alice's and Patti's work on the least common multiple. Patti's ability to carry on the inferential reasoning typical of Stage III is supported by her object conception of multiple; her confident use of the properties of multiples allows her to infer the structure of the prime factorization of the LCM. Alice, on the other hand, is unable to make similar inferences because the comparisons needed to construct common multiples require actions on multiples. Since she does not appear to have an object conception of multiple, she is led to a disequilibrating attempt to apply actions to processes. An additional example of the importance of object conceptions in inferential reasoning was seen in Karyl's work. It seems that the growth in her ability to view the factors embedded within prime factorizations as objects in their own right results in her correct inferences about divisibility that, eventually, did not need to be tested through action.

Pedagogical Suggestions

As teacher educators, we have an interest in the practical as well as theoretical aspects of understanding the nature of prospective elementary teachers' knowledge about divisibility, and of multiplicative structure in general. In particular, the APOS analyses given in the cases suggest to us some pedagogical hypotheses that we now present.

Emphasizing the Central Role of Multiplication

We first reconsider the recommendation of Zazkis and Campbell (1996a, p. 562) concerning the roles of both division and multiplication in divisibility issues. Our analysis suggests that it is mainly the role of multiplication that needs more emphasis. While one effect of this emphasis would be to reduce the primacy of division in students' thinking about these tasks, another goal would be to strengthen students' understanding of division in its role as the reversal of the process of multiplication. The act of reversing a process and reflection on that reversal are important parts of understanding a process. Reflecting on a process is required for encapsulating that process to an object, and the status of multiplication and division as objects of thought is essential to prospective teachers' knowledge of elementary mathematics.

We therefore suggest strong emphasis on the role of multiplication when introducing divisibility concepts. Explicit attention should be given to the equivalence of these relations for natural numbers A and B: "A is divisible by B," "A is a multiple of B," "B is a factor of A," and "B is a divisor of A." Students should be asked to interpret a variety of situations in terms of each concept, in both decimal and prime-factored form. Exercises should highlight the fact that the essence of all four statements is the same—simply that there exists a natural number that can be multiplied by B to obtain A.

Constructing Actions and Processes on Prime Factorizations

In order to use multiplicative structure in reasoning, we have seen that one needs to deal flexibly with prime factorizations, and to be able to identify multiples and factors as well as divisors and quotients through an application of the processes underlying the commutative and associative properties of multiplication. A major goal of experiences of this type would be not only to teach students the procedures (actions) needed for dealing with numbers in prime-factored form, but to get students to reflect on these actions sufficiently and in productive enough ways that they are able to move beyond successful actions to processes and objects.

There is one prominent feature of our data (and also reported by Zazkis & Campbell, 1996a) that is more prevalent than is suggested by the case studies we presented. We refer to the tendency of many students to return numbers in prime-factored form to decimal form in order to make decisions about divisibility. We believe that it is common for students to see prime factorizations as processes. Rather than treating a number in prime-factored form as a single number, students often perceive it only as a string of numbers, providing a set of instructions for obtaining the intended number. When reasoning about this number is required, the tendency is to carry out the operations indicated by the notation, thereby returning the number to decimal form. Flexibility in acting on and thinking about prime factorizations might be encouraged by having students construct actions on prime factorizations because, according to APOS theory, these actions, together with reflection upon the actions and their results, should lead to the encapsulation of the prime factorization as a cognitive object. The actions we suggest include arithmetic actions such as multiplying or dividing prime factorizations to obtain other prime factorizations, and applying the processes underlying the commutative and associative properties of multiplication to derive factorizations involving composite factors. Explicit attention should be focused on having students explain why their typical strategies of looking for expressions that are "in there" are reasonable, in terms of the multiplicative structure of the numbers and the operations that are being used.

Linking LCMs across Representations

In decimal form, the "create a multiple and divide" method is an easy-to-use and easy-to-justify method of constructing a least common multiple. Although it was not presented in class, several of the interview participants used it spontaneously in their solutions during the interview. This method can readily be reconstructed to a method for finding LCMs in prime-factored form. Through the reconstruction, we conjecture that students can develop an understanding of the rationale for the "higher exponent" rule for LCMs.

Briefly, the extension of this method to find LCM(A,B) in prime-factored form goes as follows: one starts with the prime factorization of A and then multiplies, sequentially, by the prime factors of B that are "missing." This has the benefit of emphasizing that a sequence of multiples of A is being created through the process of multiplication, and that the goal is to multiply by only those factors needed to obtain a number that is obviously divisible by B.

That this procedure produces the least common multiple is immediate, assuming an acceptance of the uniqueness of prime factorizations that is guaranteed by the Fundamental Theorem of Arithmetic. We conjecture that problems concentrating on the process of reconstruction may also decrease the likelihood of the construction of separate conceptions of LCM in different situations, such as we saw in Adam's work. Once the new method is understood, reflection on that process can lead students to infer the "higher exponent" rule.

Encouraging Mathematical Behavior

We believe, as do many other researchers and educators, that in order to improve the teaching of elementary school mathematics, it is critical to improve the mathematical knowledge of teachers. Liping Ma (1999) provides evidence that what distinguishes the more effective teachers of elementary mathematics from those who are less effective is the degree to which they have a "profound understanding of fundamental mathematics" (PUFM), where the term "profound" connotes both depth and breadth of understanding. As Ma defines it, understanding a topic with depth requires relating that topic to the most conceptually powerful and central ideas in the domain, while understanding a topic with breadth requires relating that topic with those of similar and less conceptual power. It is this thorough understanding, with both depth and breadth, that gives mathematical knowledge a coherent structure. Restated in terms of APOS, those with PUFM have thematized schemas for the fundamental concepts of arithmetic.

An important factor in developing PUFM in prospective elementary teachers is maintaining a classroom culture that encourages mathematical discourse and exploration. Ma noted that teachers with PUFM are also those who "tend to think rigorously, tend to use mathematical terms to discuss the topic and tend to justify their opinions with mathematical arguments" (1999, p. 105). We see this as another description of the properties of mathematical behavior that begin to appear in Stages II and III of the stage model used in our study. In our interviews, we saw examples of the type of behavior we hope to encourage, such as:

- Alice's action-oriented confirmation of a hypothesis about divisibility in prime-factored form, followed by her exclamation, "Well, let's find out why!"
- Karyl's conscious and extended efforts at trial and error in testing her hypotheses about what might be true in prime-factored form by comparing it to what she knows is true in decimal form, as well as her eventual conscious shift to inferential reasoning.
- Patti's deliberate reasoning about least common multiples that included attention to operations, and an explication of both necessary and sufficient conditions on her algorithm.

As in Ma's study, these were responses to an interviewer's probe for evidence that the individual understood the rationale for the procedures he or she had used successfully. Since success with procedures can occur in the absence of conceptual understanding, future teachers must experience situations in which their rationales for the procedures of arithmetic are revealed and challenged, enhanced, or affirmed. Opportunities to reconstruct their conceptions, with the aim of developing rationales that are explicitly expressed in mathematical terms, can help them begin to build the essential content knowledge needed for competence in teaching elementary mathematics.

CONCLUSION

We have found our theoretical approach to be helpful in describing and explaining the progress that prospective elementary teachers can make in applying their knowledge of multiplicative structure to divisibility tasks. It has helped us lay a foundation for future studies of the divisibility schema and has also led to some specific pedagogical suggestions. We acknowledge that there is still work to be done in investigating how to help prospective elementary teachers make the progression from actions, through explorations with operations, to conscious and deliberate reasoning that draws upon a mature schema for multiplicative structure.

NOTES

1. The study reported in this chapter was partially supported by a grant from the ExxonMobil Education Foundation. Also, the authors would like to thank the members of RUMEC (Research in Undergraduate Mathematics Education Community) for their insightful comments on earlier drafts of this chapter.

2. Editors' Note: The process of linking, or connecting, various actions, processes, and objects as a schema is basically the sense in which other authors in this volume are using the term "thematization."

3. Piaget's use of the word "action" in this setting generally corresponds to our use, in APOS theory, of both of the terms "action" and "process": our understanding is that his use of the word "action" may also refer to interiorized actions, which are referred to as processes in APOS theory.

4. To maintain a distinction between cognitive and mathematical issues, we refer to division as the *reverse* of (the process of) multiplication. We use the word "inverse" to refer to the mathematical fact that division by a number is the inverse of the (function defined by) multiplication by the same number.

5. The process that is encapsulated to give the associative property is taken to be the process of comparing the multiplication processes $a(bc)$ and $(ab)c$ to see that they are equivalent. Similarly, the process underlying the commutative property is the process of checking that the processes ab and ba are equivalent. Transforming the product $3 \times 3 \times 5 \times 7$ into the equivalent product 21×15, for example, would require several applications of these processes of comparison.

REFERENCES

Asiala, M., Brown, A., DeVries, D. J., Dubinsky, E., Mathews, D., & Thomas, K. (1996). A framework for research and development in undergraduate mathematics education. *Research in Collegiate Mathematics Education, 2*, 1–32.

Clark, J., Cordero, F., Cottrill, J., Czarnocha, B., DeVries, D. J., St. John, D., Tolias, G., & Vidakovic, D. (1997). Constructing a schema: The case of the chain rule. *Journal of Mathematical Behavior, 16*(4), 345–364.

Clark, J. M., Dubinsky, E., Loch, S., McDonald, M., Merkovsky, R., & Weller, K. (2000) *An examination of student performance data in recent RUMEC studies*. Manuscript submitted for publication.

Dubinsky, E., & McDonald, M. (in press). *APOS: A constructivist theory of learning in undergraduate mathematics education research.* In D. Holton & M. Niss (Eds.),

Teaching and learning mathematics at the university level—An ICMI study (pp. 273–280). Dordrecht: Kluwer Academic Publishers.

Freudenthal, H. (1983). *Didactical phenomenology of mathematical structures*. Dordrecht: Reidel.

Ma, L. (1999). *Knowing and teaching elementary mathematics*. Mahwah, NJ: Erlbaum.

Piaget, J. (1978). *Success and understanding*. Cambridge, MA: Harvard University Press.

Sierpinska, A. (1994). *Understanding in mathematics*. London: Falmer Press.

Zazkis, R., & Campbell, S. (1996a). Divisibility and multiplicative structure of natural numbers: Preservice teachers' understanding. *Journal for Research in Mathematics Education, 27*(5), 540–563.

Zazkis, R., & Campbell, S. (1996b). Prime decomposition: Understanding uniqueness. *Journal of Mathematical Behavior, 15*(2), 207–218.

4

Language of Number Theory: Metaphor and Rigor

Rina Zazkis

The use of mathematical language by learners and teachers of mathematics has attracted the interest of many educators. Researchers have pointed to problematic application of verbal cues, such as *same, altogether*, or *left* (Nesher & Teubal, 1975) and have acknowledged lexical ambiguity between the mathematics usage of words, such as *volume, product*, or *combination* and their everyday usage (Durkin & Shire, 1991). There has been a call to avoid the use of certain words, such as *reduce* or *cancel*, as they fail to describe accurately mathematical operations (Hanselman, 1997). There are a variety of reports from instructors on emphasizing mathematical language at the college level (Esty & Teppo, 1994) as well as at the elementary school level (Buschman, 1995; Cramer & Karnowski, 1995). There is an increasing awareness of the importance of communication and mathematical discussion in the mathematics classroom (Hoyles, Sutherland, & Healy, 1991; Sfard, Nesher, Streefland, Cobb, & Mason, 1998). Furthermore, the new *Principles and Standards for School Mathematics* consider students' ability to "use the language of mathematics to express mathematical ideas *precisely*" (emphasis added, National Council of Teachers of Mathematics, 2000, p. 63) as one of the instructional goals for all grade levels.

However, despite significant attention to specific words used and misused in particular mathematical topics, words of elementary number theory were not mentioned among the numerous examples. Recent research on learning number theory concepts (Zazkis, 1998a, 1998b; Zazkis & Campbell, 1996a, 1996b) created the necessity to attend to language used in these topics.

The gap between the formal mathematical language and the form of language preferred by students has been mentioned in earlier research (Zazkis, 2000). It reported a teaching technique that proved useful in helping students acquire the formal mathematical language. This report focuses on the description of the phenomenon, and seeks for causes, rather than for "cure." It attempts to identify the psychological as well as linguistic reasons that influence students' language.

DATA SOURCE

As part of a research project on learning elementary number theory by pre-service elementary school teachers, a variety of clinical interviews were conducted over a period of four years. Participants in these interviews were undergraduate students enrolled in a course called "Principles of Mathematics for Teachers," which is a core mathematics content course in a teacher education program for certification at the elementary level. There were between sixty and eighty students enrolled in each offering of the course, approximately twenty students in each cohort volunteered to participate in clinical interviews.

The interview questionnaires were designed as a part of ongoing research; the questions investigated participants' learning and understanding of basic concepts of number theory, such as divisibility, prime decomposition, factors, divisors, and multiples. Students' use of language was not a focus of investigations when the questions were designed and interviews were conducted.

MATHEMATICAL VOCABULARY

Divisibility is one of the fundamental concepts in number theory that can be captured with a variety of lexical descriptions. In a formal mathematical definition we say that a whole number A is divisible by a whole (nonzero) number B if and only if there exists a whole number C such that $A = B \times C$. (For simplicity I restrict the discussion here to the set of whole numbers, though analogous definitions may refer to integer numbers as well.) This definition is interpreted in some texts for elementary school students and teachers in terms of division: If the result of division of A by B is a whole number, then we say that A is divisible by B. There are five mathematically equivalent statements that express the idea of divisibility:

1. A is divisible by B
2. B divides A
3. B is a factor of A
4. B is a divisor of A
5. A is a multiple of B

I will refer to the use of these expressions as utilizing formal mathematical vocabulary. These terms and expressions were familiar to students as they were defined in the textbook and used in the class work for approximately four weeks of attending to the topic on number theory in the course. All the interviews took place upon the completion of the unit on number theory.

INFORMAL VOCABULARY

Considering the transcripts of the interviews it is evident that the language used by the interviewers was different from the language used by the interviewees. Most participants do not use formal mathematical terminology at all or use it inconsistently, that is, they use formal and informal terminology interchangeably. Below is an example of an interchange between an interviewer and Bob. The interviewer asked Bob to consider the number $M = 3^3 \times 5^2 \times 7$ and determine divisibility of this number by 7, 5, 2, 9, 81, 63, and 15. Bob's ability to recognize the divisibility of M by prime numbers and his difficulties and hesitation regarding the divisibility of M by composite numbers are discussed in detail in Zazkis and Campbell (1996a). I would like to focus here on the language pertaining to divisibility that is used by the interviewer and by Bob.

Interviewer: Bob, we're going to start off, I'm going to ask you to write down a number please. And that number is $3^3 \times 5^2 \times 7$, and we're going to call this number M. Now, my first question is, **is M divisible by 7**?

Bob: Yes, it is.

Interviewer: And would you explain why?

Bob: Well if 7 [*pause*], let's see [*laugh*], M is, or let's see, so **7 is a factor of M**, therefore, **it's divisible by** M, pardon me, by **7**.

Interviewer: And how about 5?

Bob: **Five is also a factor of M**.

Interviewer: Okay, and would **M be divisible by 2**?

Bob: No, it would not, since 2 is um, [*pause*] since 2 is not seen here, **it's not a factor of M**.

Interviewer: Hmm, okay, and uh why do you feel that that's the case?

Bob: Um, explain this clearly [*pause*], since 2 is not one of the numbers that's being multiplied, the **product therefore, cannot be divided by 2**.

Bob: Okay. And that since obviously 2 is a prime number, the prime number of 2 is not in this solution, therefore, whatever the product M turns out to be, **2 cannot divide into that**.

Interviewer: Um hm, okay. How about 9? **Would M be divisible by 9**?

Bob:	[*pause*] Yes it could because, [*pause*] 3^3, pardon me, **3 can go into, is a, is a factor of 9**, so any **multiple of 3 could go into that, into M**, as long as it's not greater than M.
Interviewer:	Okay. How about um let's say 81? Would you think that **81 would divide M**?
Bob:	I'd want to find out what M would be, I guess that's the, the best thing, that's what I'd prefer.
Interviewer:	Um hm.
Bob:	I guess knowing what M would equal, and then from there working backward, finding **which numbers can go into that**.
Interviewer:	Um hm.
Bob:	Um, right now I can't see whether or not **81 can go in there**.
Interviewer:	OK. Uh, how about 63?
Bob:	[*pause*] Once again, um, we have 7 now, **7 can go into 63**, well, 3 can as well [*pause*], once again I'd have to solve for M, in order to find out whether **63 can divide M**.
Interviewer:	Okay, so when you say solve for M, you mean multiply it out and then divide by 63?
Bob:	Yeah, exactly, exactly.
Interviewer:	OK. How about if you wish **to divide M by 15**?
Bob:	[*pause*] Um, well since there's 5, 5^2 in this problem, we know that the, that the units digit will be 5, now 15 obviously has a 5 in it as well, therefore quite possibly **15 will go into M**, and once again I'd have to solve for that.

In this conversation we notice that Bob starts by utilizing formal mathematical vocabulary claiming that "7 is a factor of M" as well as "5 is also a factor of M." He correctly claims that 2 is "not a factor of M." However, when asked to explain his decision, he expresses his justification by saying "cannot be divided by 2" and further "2 cannot divide into that," both expressions in this case are examples of informal terminology. When asked about the divisibility of M by 9, Bob says "3 can go into, is a, is a factor of 9, so any multiple of 3 could go into that, into M," which appears as Bob made an attempt to correct his language, but then gave up. He continues to use informal expression "can go into" when asked to determine divisibility of M by 81, 63, and 15. It appears as though Bob's language deteriorates together with his level of confidence. Although, he manages to use a formal expression "63 can divide M" among several informal ones.

Analyzing students responses and their descriptions of divisibility, four different not disjoint themes in application of informal vocabulary were identified: (a) attempt to interpret the word *divisible*, (b) invoking images and processes, (c) seeking confirmation of meaning, and (d) overemphasizing. In what follows, I describe and exemplify each theme. Further, I describe unifying features in these themes in an attempt to understand the forces that guide participants' choice of words.

Interpreting the Word *Divisible*

It is a common phenomenon that students who are not familiar with a mathematical meaning of a given word assign to it meanings from everyday situations. Pimm (1987, pp. 84–85) described an enlightening example of a child's interpretation of the word *diagonal*. A variety of other examples are mentioned by Durkin and Shire (1991).

Divisible is not commonly used in everyday situations. However, the structure of the word invites interpretation consistent with the meaning of other adjectives derived from verbs using the suffix "-able" or "-ible." For example, "eatable" means edible, or "can be eaten"; "breakable" means fragile, or "can be broken"; "workable" means feasible, or "can be worked out"; "defensible" means justifiable, or "can be defended." Using this structural analogy it is reasonable to interpret "divisible" as "can be divided."

Unfortunately, when this analogy is used in a mathematics context of number theory, it does not communicate the meaning of *divisible* in the mathematical register of English. In fact, the claim "can be divided" does not provide any information about the numbers and the relationship between them. Thinking of division as an operation between rational numbers, any number can be divided by any (nonzero) number, and the result of this operation is a rational number. This in inherent in the field structure of rational numbers. Thinking of division as an operation between whole numbers, any number can also be divided by any (nonzero) number, and the result of this operation is a pair of whole numbers, quotient and remainder, where the remainder can be zero. This is inherent in the division algorithm, that guarantees that for any whole numbers, A and (nonzero) D, there exist unique whole numbers quotient, Q, and remainder, R, where $0 \leq R < D$, such that $A = QD + R$. (For a detailed discussion of rational number division versus whole number division, and the division algorithm, see Campbell, Chapter 2, this volume.)

In the above excerpt, Bob claimed that the number M "cannot be divided by 2" meaning that M was not divisible by 2. Kate and Marty were asked to explain the meaning of several number theory concepts. Their definition of prime numbers was as follows:

Kate: **Prime numbers**? Well, according to the definition they're **numbers that can only be divided by themselves and one** . . .

Marty: Prime number. Prime number is a number that **can only be divided by one and itself**, there is **no other factors that go into that number**.

It is reasonable to conclude that Marty and Kate understood the concept of a prime number. However, they fail to explain it using appropriate vocabulary. Jane used a similar expression in a different problem situation:

Interviewer: Alright Jane, I'm going to ask you now to um, uh, write down another number, and this one is $6 \times 147 + 6$, [. . .] Um, now if we take that number and divide it by 6, what would the remainder be?

Jane: [*pause*] I don't know, I don't think there would be a remainder.

Interviewer: OK, and why do you say that?

Jane: [*pause*] Um, because **you're timesing it by 6, it's a multiple of 6, so therefore, it can be divided by 6**.

Jane's phrase "can be divided by 6" is used to communicate the idea that the result of this division is a whole number. It is reasonable to conclude here that Jane understands that the number given by $6 \times 147 + 6$ is divisible by 6. Jane invents words such as *timesing*, then she applies appropriate vocabulary, saying, "it's a multiple of 6," however, in the same breath she uses an expression inappropriate to the situation by claiming "it can be divided by 6."

Interviewer: On your calculator you performed division and decided that 23×17 is 391. What do you think, will this number be divisible by 46? **Is 391 divisible by 46?**

Andy: Um, [*pause*] yes.

Interviewer: Can you please explain?

Andy: Um, twen . . . oh, [*pause*], maybe not.

Interviewer: My request is still the same, would you please explain?

Andy: Can I do it on my calculator?

Interviewer: Sure.

Andy: OK. [*Pause*]

Interviewer: But even if you do it on your calculator, I'll ask you to explain.

Andy: Yeah, I knew that. OK, no, **not evenly**, because um here you're doubling it to 46, 23×2 is 46, oh, wait, right, and then um to find the answer you'd have to divide this one by 2 and 17 doesn't, **can't be divided by 2**.

Andy has a good idea in her explanation. She knows that $391 = 23 \times 17$. She also noticed that $46 = 23 \times 2$. She tries to explain that one of the factors of 391 was doubled, therefore the other one has to be halved to reach the same product. However, her wording "17 can't be divided by 2" is an inaccurate choice of words to describe the situation. It is also interesting to note that the interviewer's question is asked in terms of divisibility: "Is 391 divisible by 46?" However, Andy made no attempt to use the language modeled by the interviewer. In Andy's words we recognize another recurring issue, the issue of "even" division. This issue is exemplified further in the excerpt with Jill. Jill, when asked to explore divisibility of a number by 15, replied:

Jill: Because once again I know what M is, it's 1,575, and I know that 1,500 **can be divided um by 15 evenly** and I know that 75 also **can be divided evenly by 15**.

The intended meaning of "can be divided evenly" is clear; however, the reference to even division is an avoidance of proper mathematical terminology. Anna, when asked whether 391 was divisible by 23, offered a conjecture that 391 was prime. The following is her justification for this conjecture:

Anna: Because I don't think it **had been divided** by 2 or 3, or 5 or 7 [*laugh*].

A variation on this expression is found in Barb's words, who offered the same conjecture:

Barb: I looked at that, but I couldn't find a number that would **easily be divided into 391**, and so I just assumed that it was prime, and, and, I mean I could be wrong, I didn't, I didn't take 8 or something like that, but I just checked for the first five numbers.

The tendency of some students to conclude that the number that is not divisible by the first five primes is prime is described in detail by Zazkis and Campbell (1996b). Here, I wish to focus on Barb's expression "easily be divided into 391." Can 344,578 be easily divided by 457? Can 7 be easily divided by 2? The division in the second case is easier for me. However, regardless of the ease with which this division is performed, 344,578 is divisible by 457, whereas 7 is not divisible by 2. Barb's search for factors is an appropriate activity; however, she describes this search using ambiguous and imprecise terminology.

Invoking Images and Processes

Another common theme in students' description of divisibility can be outlined as invoking a mental image or a process of division. A mathematical definition of divisibility can be interpreted in the following way: A number A is divisible by a number B if A objects can be arranged in B groups (rows, columns) such that there is the same number of objects in each group. This interpretation is consistent with a partitive view of division. Taking a quotitive or measurement view on division, the image that may accompany the division is covering the length of A units with segments B units long. In saying "goes into," "fits into," or "can be put into," the participants could be visualizing this process of division. In fact, "goes into" or "goes into evenly" appeared to be the most popular expression used to describe divisibility by preservice teachers in this study.

Interviewer: How many **divisors** would you expect the number 90 to have?

Barb:	Would just expect, like, expect from it, or do you mean like, should I actually figure it out?
Interviewer:	Well, what you could do is, you can tell me how you would go about determining the number of **factors** that 90 has.
Barb:	Okay. Well, um, **1 goes into it, then 2 goes into it, 3 goes into it, um, 4 doesn't go into it, 5 goes into, 6 goes into it**, 7 doesn't, 8 doesn't, 9 does, 10 does.

Note that Barb considers the numbers that "go into it" following the question about divisors or factors of the number 90. The interviewer's language is clear to Barb, but she avoids using it.

The following interchange takes place as the interviewer discussed with Darlene the number $A = 6 \times 147 + 1$:

Interviewer:	Alright. Now, I'm going to ask you to find the smallest number that's bigger than A that's **divisible by 6**.
Darlene:	Okay. Well, uh, 889, no, I don't know [*laugh*], it would be 888.
Interviewer:	You've gone back to your number, and you've added something to it.
Darlene:	Yes, I added 6 to 883, no I added 6 to 882, because I know **6 can be put into 882, so 6 will fit into 882 + 6**.

Darlene's strategy to find the smallest number that is bigger than A and divisible by 6 is to consider the closest multiple of 6, and add 6 to it. Her metaphorical language invokes the process of measurement division and her explanation avoids the reference to "divisible by 6," as modeled by the interviewer, instead referring to "can be put into" and "will fit into."

In the next excerpt the interviewer asks Barb whether 391 is divisible by 46. In her answer, Barb uses formal and informal expressions interchangeably.

Interviewer:	I guess the next question is, **would 391 be divisible by 46**?
Barb:	[*Pause*] I would imagine so, but, I guess I should work it out again. [*Pause*] Oh my goodness, I'm sorry, my previous answer is wrong, but I'm just, figure out this one, [*long pause*] no, it's not [. . .] 46, it seemed natural that it would, that 391 would, or would **be divisible by 46**, as **46 is a multiple of 23**, but it doesn't **go in evenly**.

Marty was asked to describe how she would determine all the factors of a given number. Marty's expression "go into it completely" is a variation on the more popular expression of "go in evenly."

Interviewer:	How would you go about listing all of the factors of 117?

Marty: Probably the same way, as **dividing the numbers into it and seeing if they go into it completely** and then if they do, then that's all, that's the factors.

Students' strong procedural dispositions have been identified and described in Zazkis and Campbell (1996a). The majority of participants considered divisibility as a consequence of performing division, rather than as a relationship among natural numbers. These pervasive dispositions toward procedural attachments are supported by the language students chose to use—the language that describes the process.

Seeking Confirmation for Meaning

On several occasions, such as in the excerpts below, students applied at times informal vocabulary to confirm for themselves the meaning of the word intended by the interviewer. These concepts were used in class a month prior to the interviews as students were solving problems pertaining to elementary number theory. All interviewees had experienced tasks in which they were asked to determine the divisibility of and by specific numbers. However, a request for interpretation is a student's desire to establish a shared meaning, a request for the interviewer to confirm the meaning assigned by the student.

Interviewer: Let's go back to our number M. Do you think **M is divisible by 63**?

Sandra: You mean whether 63 will **fit into this number, without remainder** you mean?

Interviewer: How about 15, if you were to divide M by 15, what do you think the remainder would be of that division?

Darlene: I don't know. Are you asking **if it would go in evenly**?

Interviewer: Is 391 **divisible** by 23?

Jane: **Divisible evenly**?

Interviewer: If you were to establish for yourself, one way or another, whether or not **23 divides 391**, how would you go about it?

Laura: [*pause*] Do you want it to **divide evenly, with no decimals**?

It is also interesting to note that such rephrasing for the sake of clarification and confirmation did not occur when the interviewer used the concept for the first time. It appeared when the interviewee was unsure in her answer. For example, Sandra did not seek clarification when asked about divisibility of M by 7 or 5. She did seek clarification when a factor was not immediately recognized and therefore performing division appeared to be her means of drawing a conclusion about divisibility.

"Evenly," as a connection between divisibility and division, can be interpreted as "either the fractional component of the rational quotient or the whole number remainder must be zero" (Campbell, Chapter 2, this volume). Further, Campbell suggests that many of the difficulties that preservice teachers experience with arithmetic division may be due to their "lack of discernment regarding the numerical domains." Participants in Campbell's as well as in this study were uncertain in determining whether rational number division or whole number division is intended and lacked a clear distinction between the two. The language that students use is a manifestation of this confusion. It appears that Sandra, in her reference to "without remainder," had whole number division in mind, whereas Laura, thinking of division "with no decimals" was referring to rational number division. Laura's clarification signals that whole number division, which is intended in number theory situations, was not automatically assumed. There is no explicit reference to rational or whole number division in excerpts with Jane and Darlene and the use of "evenly" can imply either. In fact, as shown in the next section, "evenly" is a popular expression used not only to clarify but also to emphasize the intended meaning.

Overemphasizing

Interviewer: OK. Let's call this number: $(6 \times 147) + 1$, let's call that number A. Can you find the smallest number that is bigger than A and **divisible** by 6?

Linda: [*Pause*] Um, **evenly divisible by 6**. [*Pause*] I would, I would add 5 to the number that's outside the brackets, making it $6 \times 147 + 6$ and, without working it out, I would assume that that would make it **evenly divisible** by 6 and be the next largest number.

The interviewer invited Linda to look for a number divisible by 6. She immediately interpreted this request as "evenly divisible." Through the interview, she continued to use the expression "evenly divisible" in places where "divisible" would be appropriate. Linda's choice of words is her way to assure the meaning of the concept. It also can be seen as a compromise between the use of "goes into evenly" utilized by many of her classmates and the use of "divisible" utilized by the interviewer. The following excerpts exemplify a similar phenomenon.

Interviewer: Is M divisible by 7?

Jack: By 7 you **should be able to divide it**, quite possibly have the potential of **complete divisibility**.

Interviewer: I'm going to ask you, is 391 divisible by 23?

Lisa: Hmm, [*pause*] um, well, [*pause*] um well, it is, but I'm not sure if it's **divisible evenly or not**.

Jill: I only have to add an extra 5 **to make that divisible evenly by 6** . . .

Lee: I think, I think **it's divisible evenly by 15**, because there's a 5 as a factor and we can multiply that by one of the 3s, is 15.

Lucy: You're asking me to find a 5-digit number, such that when divided by 2 has a remainder of 1. [*Pause*] I'm thinking it would probably have to be an odd number, because **all even numbers would be evenly divisible by 2** . . .

In the same way as students refer to *whole*, rather than *absolutely whole*, numbers or *proper*, rather than *completely proper*, fractions, it is important to learn to refer to *divisibility*, rather than *complete divisibility* or *even divisibility*. Conciseness is one of the features of mathematical vocabulary. Whenever adverbs or adjectives are added to mathematical concepts, they alter the meaning. Compare, for example, *converging sequences* with *absolutely converging sequences* or a *relation of order* with a *complete relation of order*.

The concept of divisibility already captures the meaning; however, the word *divisible* does not communicate this meaning for the students. As discussed earlier, the structure of the word *divisible* and the familiar notion of division invites interpretation of "divisible" as "can be divided." However, realization that "can be divided" is meaningless in a discussion of the multiplicative structure of numbers provokes the students to modify the usage of the word *divisible*. The meaning students wish to communicate is captured by adding unnecessary adjectives to a mathematical term.

CONCLUSION: METAPHOR AND RIGOR

During my ongoing experience of teaching mathematics for preservice elementary school teachers, I was not only troubled but also fascinated by students' reluctance to adopt formal mathematical terminology. My successful as well as unsuccessful attempts to deal with this instructional challenge are reported in Zazkis (2000). In this chapter, I attempted to explain possible reasons for students' reluctance to use formal terminology and robustness of informal expressions in their vocabulary. Four interconnected factors—use of linguistic analogy in interpreting the unfamiliar word *divisible*, invoking mental images, seeking confirmation for meaning, and overemphasizing—were outlined. However, it was Sfard's (2000) metaphor of "steering the discourse between metaphor and rigor" as a story of development of mathematical ideas that illuminated a common trend in students' language, other than being informal, imprecise, and at times inappropriate in a given situation.

According to Sfard (2000), the need for effective communication is the driving force behind students' construction of mathematical objects. Further, Sfard suggests that mathematical objects emerge through negotiations between metaphor and rigor. The virtual mathematical object of conversation in all the above examples is divisibility. However, the word *divisible*, as well as other formal expressions, even when introduced and modeled by the instructor, do not capture for participants, at least initially, the idea of divisibility. Therefore, motivated by the need to communicate, students' choice of informal expressions is initially guided by a metaphor. "Goes into" or "fits evenly" describes for students the multiplicative relationship of divisibility, and this metaphor helps students in constructing the idea of divisibility before they are able to express this idea in formal mathematical discourse.

Lave and Wenger (1991) argue that learning in a community of practice involves learning the language of this community. In particular, learning mathematics includes "appropriating ways of speaking mathematically" (Adler, 1998, p. 30). A necessary feature of successful communication is that parties involved in it refer to the same things when using the same words (Sfard, 2000). Therefore, acquiring the language of the mathematical community is critical for the purpose of successful communication. Helping students, and preservice elementary school teachers in particular, in this endeavor presents a pedagogical challenge. Approaching this challenge as "steering between metaphor and rigor" is a useful perspective.

REFERENCES

Adler, J. (1998). A language of teaching dilemmas: Unlocking the complex multilingual secondary mathematics classroom. *For the Learning of Mathematics, 18*(1), 24–33.

Buschman, L. (1995). Communicating in the language of mathematics. *Teaching Children Mathematics, 1*(6), 324–329.

Cramer, K., & Karnowski, L. (1995). The importance of informal language in representing mathematical ideas. *Teaching Children Mathematics, 1*(6), 332–335.

Durkin, K., & Shire, B. (1991). Lexical ambiguity in mathematical contexts. In K. Durkin, & B. Shire (Eds.), *Language in mathematical education: Research and practice* (pp. 71–84). Milton Keynes, UK: Open University Press.

Esty, W. W., & Teppo, A. R. (1994). A general education course emphasizing mathematical language and reasoning. *Focus on Learning Problems in Mathematics, 16*(1), 13–35.

Hanselman, C. A. (1997). Stop using foul language in the mathematics classroom. *Mathematics Teaching in the Middle School, 3*(2), 154–160.

Hoyles, C., Sutherland, R., & Healy, R. (1991). Children talking in computer environments: New insights into the role of discussion in mathematics learning. In K. Durkin & B. Shire (Eds.), *Language in mathematical education: Research and practice* (pp. 162–176). Milton Keynes, UK: Open University Press.

Lave, J., & Wenger, E. (1991). *Situated learning: Legitimate peripheral participation.* Cambridge: Cambridge University Press.

National Council of Teachers of Mathematics. (2000). *Principles and standards for school mathematics*. Reston, VA: Author.

Nesher, P., & Teubal, E. (1975). Verbal cues as an interfering factor in verbal problem solving. *Educational Studies in Mathematics, 6*, 41–51.

Pimm, D. (1987). *Speaking mathematically: Communication in mathematics classroom.* London: Routledge.

Sfard, A. (2000). Steering (dis)course between metaphors and rigor: Using focal analysis to investigate an emergence of mathematical objects. *Journal for Research in Mathematics Education, 31*(3), 296–327.

Sfard, A., Nesher, P., Streefland, L., Cobb, P., & Mason, J. (1998). Learning mathematics through conversation: Is it as good as they say? *For the Learning of Mathematics, 18*(1), 41–50.

Zazkis, R. (1998a). Odds and ends of odds and evens: An inquiry into students' understanding of even and odd numbers. *Educational Studies in Mathematics, 36*(1), 73–89.

Zazkis, R. (1998b). Divisors and quotients: Acknowledging polysemy. *For the Learning of Mathematics, 18*(3), 27–30.

Zazkis, R. (2000). Using code-switching as a tool for learning mathematical language. *For the Learning of Mathematics, 20*(3), 38–43.

Zazkis, R., & Campbell, S. (1996a). Divisibility and multiplicative structure of natural numbers: Preservice teachers' understanding. *Journal for Research in Mathematics Education, 27*(5), 540–563.

Zazkis, R., & Campbell, S. (1996b). Prime decomposition: Understanding uniqueness. *Journal of Mathematical Behavior, 15*(2), 207–218.

5

Understanding Elementary Number Theory at the Undergraduate Level: A Semiotic Approach

Pier Luigi Ferrari

Elementary number theory is a subject often disregarded in high school curricula; in some countries, such as Italy, subjects like linear algebra and elementary calculus occupy a central role, both in curricula and in examinations. As a result, most students are accustomed to working with real or rational numbers. They often do not realize that some methods that work in these domains do not work within the ring of integral numbers, nor, for that matter, in any other ring that is not a field. University curricula also tend to neglect elementary number theory. Although introductory modern algebra courses often deal with abstract notions of group, ring, and field, concepts of number theory are typically introduced only as illustrations or applications of ring theory. In this chapter I focus on elementary number theory at the undergraduate level, and describe some pedagogical opportunities it affords in the development of advanced mathematical thinking. In so doing, I focus on difficulties experienced by freshman computer science students enrolled in introductory mathematics courses.

Recent research has called into question popular opinions that ascribe deficiencies in mathematical understanding that undergraduates manifest in university to deficiencies in specific subject content knowledge from their high school curricula. On the contrary, students' improper, or unrefined, attitudes toward mathematics have been recognized as a major source of failure. In this regard, a clear and convincing account has been provided by Vinner (1997), who uses terms like "pseudo-conceptual" and "pseudo-analytical" to denote thinking or problem solving processes that fail to capture the actual or intended understanding of concepts or resolution of problems. Difficulties with language are another major source of failure. This holds for both algebraic formalism (interpretation, translation, and transformation of formulas, etc.) and ordinary language (as it is used in doing mathematics, for example, when interpreting the statement of problems or planning a resolution strategy). Troubles with language are closely linked to students' improper attitudes as well: in problem solving, on the one hand, people adopting pseudo-analytical strategies typically bind themselves to a search of keywords or verbal clues suggesting what they are expected to do; on the other hand, people with a poor mastery of language are forced to adopt pseudo-analytical strategies, as they cannot reconstruct the meaning involved in problem statements.

Thus, improper attitudes toward mathematics and poor mastery of language may prevent students from improving their mathematical competence. Improper attitudes often induce students to try to obtain credit for mathematics without actually understanding it. Linguistic difficulties can disrupt effective communication between students and teachers, students and knowledge (books, etc.), and among students themselves. These reflections suggest that an introductory or bridging algebra course should have students focus on meanings, thus dissuading the application of algorithms by rote, and offer suitable opportunities for checking the meaning of the mathematical expressions involved.

Requirements like these are satisfied by topics from elementary number theory, such as divisibility and prime decomposition, which offer a number of pedagogical opportunities to help alleviate students' improper attitudes toward mathematics at the beginning of Introductory Modern Algebra or Discrete Mathematics courses. First, they do not require a wide theoretical background, as they are grounded on ideas and methods usually learned in primary or middle school. This aspect of elementary number theory is important because students sometimes regard deficiencies in other subjects (like trigonometry, logarithms, and so on) in their high school curriculum as the main factor of their failure (see Zan, 1997, for a discussion on the influence of factors like these in mathematical learning). Besides, the relatively small amount of conceptual prerequisites in elementary number theory provides good opportunities for actually grasping the meanings involved, through exploration, different representations, and different heuristic methods, including trial-and-error, numerical checking of results, and so on.

Second, elementary number theory involves some properties and algorithms different from those stressed in high school curricula and practice and, in particular, provides a choice of simple problems that cannot be solved by direct application of algorithms but require some interpretation of the data. This is typical of impredicative problems (i.e., problems whose unknowns cannot be explicitly represented as functions of the data). For instance, a simple problem like "Find an integer x such that $\text{GCD}(x,6) = 2$" requires nothing more than the knowledge of the meaning of GCD and of some method to compute it; it cannot be solved by means of algebraic transformations alone because it has many solutions and x cannot be represented as an algebraic function of 6 and 2. In other words, in elementary number theory there is a wide choice of problems that will force students to use knowledge and heuristics and discourage them from applying algorithms with no reference to meanings.

THE LANGUAGE OF MATHEMATICS

In this section some ideas about language and mathematical vocabulary are presented that may be useful to better understand the rest of this chapter. Suggestions have been gathered from different books and papers, both in semiotics and mathematics education. In particular, some reflections of Morgan (1998), Clark (1992), and Grice (1975) as well as the detailed framework presented by Duval (1995) have been very useful.

Some recent studies on the interactions between mathematics and language point out that it is improper to speak of "the language of mathematics" as it were univocally determined. On the contrary, there is a wide range of registers (in the sense of Halliday, quoted by Pimm, 1991)[1] that are commonly used in doing mathematics. Mathematical texts, either written or oral, may be used for different purposes (teaching, learning, popularizing, communicating, playing, . . .) by different groups (pupils and students of any age level, researchers, people from different countries with different levels of linguistic or mathematical competence, etc.).

Most mathematical registers are based on various natural languages, from which they widely borrow forms and structures, and may include both a symbolic component and a visual one. Nevertheless, in mathematical registers the use of some words and structures borrowed from natural language is a bit different from the standard one. The difference is unnoticeable in children's mathematics, whose forms and meanings have been almost completely assimilated into natural language, but it grows more and more manifest in advanced mathematics.

In more advanced mathematical registers, some words take a meaning that differs from natural language registers by adding some new technical meaning to the standard ones. This is the case for words like "power," "root," "prime," or "function." The meaning of some connectives may change too, which implies that the meaning of some complex sentences (like conditionals) may significantly differ from standard usage. Ordinary language and mathematical language registers may also differ as far as the purposes or relevance of a statement are concerned. For example, in most ordinary registers a statement like "That shape is a rectangle" *implicates* that it is not a square as well, for if it were, the word "square" would have been used as more appropriate. This additional information is called an *implicature* of the statement and is not conveyed by its content only but also by the fact that it has been uttered (or written) under certain given conditions. Also, a statement like "2 is less or equal than 10^9" is hardly acceptable in ordinary registers (as it is more complex than "2 is less than 10^9"

and conveys no more information), whereas it may be perfectly appropriate in mathematical reasoning exploiting some properties of $\leq$. A relevant source of trouble is the interpretation of mathematical language according to conversational schemes predicated on natural language. For more examples of this regard, see Ferrari (1999).

Within the wide area of study usually referred to as "mathematical language," there are a variety of notations (including words, algebraic symbolism, geometrical figures, and graphs) to represent a variety of concepts, properties, and processes. Duval (1995) names some of these notations a *semiotic representation system*.[2] According to Duval, the main semiotic activities are the *formation* of a representation, its *treatment* within a system, and the *conversion* between different systems. In Duval's opinion, the mastery of conversion processes and the ability of using different representations in the same process of thought (which he names *coordination of systems* ["coordination de registres"]) are fundamental steps to conceptualization.

Duval (1995) strongly objects to teaching methods not aimed toward the coordination of semiotic systems, but using only (or almost only)[3] one of them. A thorough discussion of these ideas is beyond the scope of this chapter. However, I do use some of these ideas to make explicit what I mean by "semiotic control," a construct I have introduced to interpret students' behaviors and that has proved extremely useful in assessing their performances. I then highlight educational implications with regard to some of the potential pedagogical roles of elementary number theory at the undergraduate level with respect to their understanding of mathematics.

SEMIOTIC CONTROL

By "semiotic control" I mean the ability to properly interpret and handle symbolic expressions and statements involved in a mathematical task as texts to be interpreted, not as carriers of keywords or any sort of clues. In the case of problem solving this means recognizing that the main purpose of statements (including those in symbolic form) is to describe problem situations rather than to prompt specific behaviors or procedures. Semiotic control is also related (but not always equivalent) to some ability to explain one's own resolution strategy; however, in this chapter I am not focusing on students' explanations. All the examples I am giving concern evaluation tests given to different groups of first-year undergraduate computer science students.

The definition of semiotic control is virtually useless without some criteria to evaluate students' behaviors. The criteria should consist in assessing subjects' behaviors in situations where pseudo-analytical strategies are not easily applicable, and the consideration of meanings is required to some extent to devise a strategy. This may happen, for example, if the selection of a strategy or its applicability depend upon some parameter occurring in the statement of the task, or if the task involves representations different from the ones that subjects are accustomed to working within, or if the initial data do not immediately satisfy the conditions required, but some adjustments are needed. Conditions such as these can quite often be readily met in problems from elementary number theory, which serve well to illustrate some of those more meaningful strategies.

Applicability of Strategies

First of all, students should be able to verify the conditions that allow the application of any method before or while actually applying it and recognize when a given method is not applicable. Let's consider some examples:

> For which integral values of m does the equation $mx + 3y = m$ have integral solutions? (1)

It is common to see students writing $y = \frac{m(1 - x)}{3}$ and then finding non-integral solutions—thus imposing wrong conditions on m, or getting stuck. This phenomenon happens even after they have been taught the theory and algorithms appropriate to handle and solve linear Diophantine equations, and may depend, at least to some extent, upon the widespread practice of solving linear equations only within a field. Here is a case in which students apply a technique utterly disregarding the conditions; the pseudo-analytic approach they take is little more than "symbol pushing": a physical manipulation, treating symbols like manipulatives rather than expressions of a semiotic system equipped with definite meanings and properties.

Let's consider two more problems:

> Consider the function f: N × N → N, defined by $f(m,n) = 2^m3^n$. Would you agree that f is 1 − 1? If you do, prove it, otherwise refute it. (2)

> Consider the function f: N × N → N, defined by $f(m,n) = 2^m4^n$. Would you agree that f is 1−1? If you do, prove it, otherwise refute it. (3)

The statements of problems (2) and (3) differ only in one number. In (2) one has just to verify that the formula

$$2^m3^n = 2^x3^y \Rightarrow (m,n) = (x,y)$$

is true, which follows from the fundamental theorem of arithmetic that establishes the uniqueness of prime decomposition. In other words, there is a theorem that allows the derivation of

$$(m,n) = (x,y)$$

from

$$2^m 3^n = 2^x 3^y.$$

When dealing with problem (3), even after seeing a worked-out version of (2) and discussing it, some students would apply the solution procedure of (2) as well. They do not seem to give any value to the reasons allowing them to apply such strategy in problem (2), nor to the knowledge used to get the solution.

Generally, problems like (3) are properly solved by more than 60 percent of the students in my classes. From a procedural perspective, (3) seems not very hard, as its complexity (partly caused by the occurrence of two variables) can be reduced by evaluating (m,n) and searching for a different pair (x, y) such that $2^m 4^n = 2^x 4^y$, or even $m + 2n = x + 2y$. This procedure leads to an answer almost easily, if one does not take $(m,n) = (0,0)$ or $(m,n) = (1,0)$ as starting values. Trial-and-error strategies might be used and some standard properties of the exponential function could make the search more effective. The key to solve the problem is to recognize that the procedure used for (2) does not work anymore. But all this requires students to deal with the meanings of the statements rather than searching for keywords suggesting what they are required to do. Students applying the same strategy used in problem (2) to problem (3) show a very low level of semiotic control. Moreover, this implies that the proper solution of (2) alone, even if equipped with some words of explanation, does not mean that students have clearly understood the mathematical reasons why their answers are correct. Therefore, evaluation practices that only ask students about problem situations, for which methods they have already learned are applicable, do not give much information about their actual understanding.

Invariance through Change of Representation

Another criterion that can be helpful for evaluating students' behavior concerns their ability to work with different representations of the same concept, to recognize the invariance of some properties, and to apply their knowledge and methods to more than one representation. This criterion has to be taken with some caution: On the one hand, it is difficult even for professional mathematicians to work with representation systems they are not accustomed to; on the other hand, some students may attach their knowledge to some representations and be unable to apply that knowledge to others, even if those others are familiar to them. Let us consider some examples.

Find a decimal number x such that $0.75 < x < 0.80$. (4)

Find a fraction $\frac{m}{n}$ such that $\frac{3}{4} < \frac{m}{n} < \frac{4}{5}$. (5)

Usually, when given to freshman students at the beginning of the initial bridging course, problems like (5) are properly solved by less than 60 percent of each sample, whereas problems like (4) are solved by more than 90 percent. If prompted, almost all the students would show a sufficient knowledge of the algorithm converting a fraction into an equivalent decimal number and the heuristic methods for converting a finite decimal number into a fraction. Yet typically, one student out of three cannot apply these methods to get an answer. By the way, the most popular wrong answers to (5) are:

> "A fraction like that does not exist because there are no integers between 3 and 4"
> "A fraction like that does not exist because there are no integers between 15 and 16"
> $\left(\text{after transforming } \frac{3}{4} \text{ into } \frac{15}{20} \text{ and } \frac{4}{5} \text{ into } \frac{16}{20}\right)$.

The percentage of students choosing these answers may range from 0 to 12 percent according to the sample and the year. Anyway, they show a poor ability to handle fractions as representations, perhaps induced by the custom of converting all data into decimal numbers and applying algorithms to them. The second answer is particularly interesting because these students show some ability to handle fractions, but do not iterate upon their procedure, which would seem the next obvious step and one that would allow them to easily find an answer. Perhaps they are applying the proper school algorithm in a stereotyped way (yet another pseudo-analytical strategy) rather than as a means to achieve their goals (such as repeating the procedure with another multiple of the denominators as a new common denominator).

In my opinion students failing to give a proper answer to (5) do not display a satisfactory level of semiotic control on the representations of rational numbers. I do not question that problem (5) may be more difficult and require more mathematical competence than (4), but since students have spent a lot of time in middle school doing exercises on fractions, it is reasonable to expect that they would have mastered the idea of fractions as a way of representing a rational number, which can be compared or combined with other representations. Students' difficulties, or so it would seem, might be ascribed to educational practices not focusing on meanings but encouraging the application of algorithms detached from specific context and goals. This does not mean that the knowledge of fractions should be a condition for semiotic control on rational numbers in any particular curriculum, but the study of fractions without semiotic control is useless.

How to introduce fractions in order to avoid the loss of semiotic control is beyond the scope of this chapter. In my opinion, any approach should focus on fractions and (finite or periodic) decimals as alternative notations for the same class of numbers more than as different concepts. Also, more explicit visual representations for fractions, like those presented by Chiappini and Molinari (2000) and based on Thales's theorem, should be helpful.

Other Criteria

There are other criteria, too, that may point to semiotic control. One is the ability to explain one's own procedures in words and to add details on request. An explanation in itself, however, may not be enough. Learners often memorize and recount explanations without ever having appreciated or mastered their meanings. A lack of accuracy, on the other hand, is not necessarily a symptom of lack of semiotic control, if the subject can improve his or her explanations on request. Another criterion is the ability to deal with complexity, that is, to apply knowledge to problem situations properly even if they are a bit more complex than usual. Let's consider some examples.

> Build a formula that for any positive integer n gives the value of the sum of all positive integers less or equal to $2n$. (6)

> Build a formula that for any positive integer n gives the value of the sum of all positive integers between n and $2n$ (including the extremes). (7)

Both problems have been given as an application of the theorem on the sum of the first n positive integers. I have found that problem (6) is typically solved by a much larger number of students than (7). Indeed, students can usually be divided into three groups of about the same size: Students falling in one group cannot make anything of writings like

$$(*) \sum_{i=1}^{n} i = \frac{n \cdot (n+1)}{2}$$

and get stuck in both problems. Some others form an intermediate group. They realize that they can simply replace any occurrence of n with $2n$ and properly solve problem (6), but not (7). Notice that this replacement does not imply a full mastery of the symbolic expressions involved, as there are students that properly solve (6) but provide formulas like

$$\sum_{i=n}^{2n} i = \frac{2n \cdot (2n+n)}{2}$$

as applications of (*) to solve (7), showing a poor mastery of the meanings of the parameters involved in the sum.

The third group tended to build a formula like:

$$\sum_{i=n}^{2n} i = \sum_{i=1}^{2n} i - \sum_{i=1}^{n-1} i$$

showing a sufficient mastery of the meanings of the expressions involved, even if their solutions are not always in closed form, or include some miscalculation. In problem (7), which surely is more demanding than (6), there are very few opportunities of finding a proper solution without considering the meaning of the expressions involved.

Generally speaking, the role of complexity is particularly intriguing from the viewpoint of semiotics: Sometimes simple statements requiring some accuracy in interpretation may be misinterpreted if included in a more complex frame. For example, a problem like:

> Find a real polynomial p such that: (8)
> (a) the degree of p is 2;
> (b) p has at least a real root;
> (c) p has at least two integral roots.

is easily solved by almost all first-year computer science students, after a few hours of work on real polynomials. The only difficulty lies in the interpretation of the words "real" and "integral" that asks for some accuracy because mathematical semantics (according to which any integer is a real as well) must be applied—rather than conversational schemes. The latter according to which "real" may mean "not integral," in the same way as "rectangle" may mean "not square."

If we include the same words in a more complex situation, like problem (9), students' behavior is much different.

> Find a real polynomial p such that: (9)
> (a) the degree of p is 4;
> (b) p has at least a real root;
> (c) p has at least two integral roots;
> (d) p has at least a complex non-real root.

Problems like this are properly solved by less than 60 percent of each sample. The main source of trouble is the failure in recognizing that any integral root is a real root as well. A fair number of students who successfully applied this property in problems like (8) seem unable to use it in this problem. Here, students' attention is focused on requirement (d), which needs the application of a theorem ("For any real polynomial p, if a complex number z is a zero of p, then its conjugate z^* is a zero as well") that students regard as important and difficult. Thus the interpretation of "real" and "integral" is not regarded as the core of the problem and is carried out by means of conversational schemes. Possibly the relative ease of (8) (in comparison with their expectations) has induced students to interpret words with some inaccuracy.

Problems like (8) or (9) are sometimes regarded as "unfair." I myself do not give them in examinations, even though I always give problems requiring some effort in the interpretation of the statements. In my opinion, it is important to compromise. On the one hand, it may be unfair to give problems like these in examinations, as they may increase students' anxiety and produce distorted results. On the other hand, stereotyped problems that do not ask for an accurate interpretation of the statements involved should be avoided, as they may encourage pseudo-analytical behaviors and promote a lack of semiotic control.

DUBINSKY'S APOS SCHEME AND SEMIOTIC CONTROL

A parallel can be drawn between the idea of semiotic control and Dubinsky's Action–Process–Object Scheme. The APOS framework (for more information, see Dubinsky, 1991; Dubinsky, Leron, Dautermann, & Zazkis, 1994; Zazkis & Campbell, 1996; or Brown, Thomas, & Tolias, Chapter 3, this volume) is a constructivist-oriented theoretical framework stating that the constructions involved in reequilibration (in the sense of Piaget) are mainly of three kinds: actions, processes, and objects. Actions are repeatable physical or mental manipulations of objects. When an action can be imagined to take place in the mind of an individual, it has been *interiorized* to become a *process*. When a process itself, as a whole, can be transformed by some action it is said to have become *encapsulated* as an *object*. The construction of connections that relate different actions, processes, or objects to a particular object is named *thematization* of the *schema* associated with that object.

Students giving wrong answers to (1), (3), or (5) (as discussed above) seemingly have not developed their constructions beyond actions, with no understanding of the influence of conditions or the opportunities offered by the notation systems (as in [5]). It is widely recognized that representations may support the process of interiorization (even though there is no complete agreement on which representations are most suitable; for a discussion, see, for example, Kaput, 1991); thus the lack of interiorization may well go together with a poor semiotic control on the natural language used in doing mathematics, including the natural language component. Also, Zazkis and Campbell argue that the process of interiorization implies "some understanding of the influence of initial conditions regarding outcome" (1996, p. 547).

The role of representations is also crucial in the transition to conceptual thinking. Here is an example:

> Find a subgroup H of the additive group of integers satisfying all of the following conditions: (10)
> (a) $8 \in H$; (b) $18 \in H$; (c) $1 \notin H$.
> How many subgroups fulfill all of the conditions (a), (b), (c)?

Example (10) can be divided into two parts. The simple construction of the subgroup is typically correctly performed by more than 60 percent of each class. From a strictly procedural point of view, one has only to apply a rule that has already been learned, that is, to compute the greatest common divisor of {8, 18} and consider the subgroup generated by it. This procedure leads to fulfilling the condition "$1 \notin H$" automatically. However, a (correct) justification regarding the uniqueness of the solution is usually given by only a small group of students. To get a proper justification, one has to suspend the previous procedure, change his or her perspective, and take into account a general representation theorem (i.e., "All the subgroups of $\mathbf{Z}$ are of the form $n\mathbf{Z}$, with $n \in \mathbf{N}$"). This representation allows students to easily handle all the (infinite) subgroups of $\mathbf{Z}$ and to go beyond a subjective interpretation of uniqueness (i.e., *"I cannot do anything else"*). The subjective interpretation of uniqueness can be accounted for within both the APOS framework and the semiotic perspective. Uniqueness (intended in the mathematical, objective sense) is a relational property (something is unique in relation to other objects), which in this case requires the activation of links with other procedures and concepts, which means that the use of knowledge (not only hands, pencil, or pocket calculator) as a tool is required. Both the lack of encapsulation and of a suitable representation of the products of procedural activities may prevent subjects from dealing with uniqueness and induce them to resort to a procedural, subjective counterpart. In the case of problem (10), the notation plays a major role both in encapsulation and thematization, as it allows a variety of algorithms to be applied to subgroups of $\mathbf{Z}$ and links the idea of subgroups with some ideas of arithmetic.

More generally, the link between representation and thematization is very close. Representations not only are necessary, for it is difficult to connect processes or objects without representing them, but also may themselves suggest some connections. Fractional notation in arithmetic, for example, which seems to be most popular among first-year undergraduates and used even to represent the divisibility relation, suggests links to the action of simplification and to properties that involve rational numbers. The notation of Euclidean division (i.e., "$m = nq + r$") may draw one's attention toward the properties of r. Thus, according to the notation adopted, a statement like "*m is not divisible by n*" might be interpreted as "$\frac{m}{n}$ is a decimal" or as "$r \neq 0$". Each of these interpretations suggests links to different mathematical settings.

Students providing correct answers to (3) and the first part of (10) seemingly can coordinate initial conditions, their actions, and the related goals, which can be regarded as a step to interiorization. Problems like the second part of (10) seem to require some form of both encapsulation and thematization, as far as abstract properties of division and the representation of a group as an object in itself (rather than as an collection of objects) are involved.

AN EXPERIMENT

The study reported in this chapter was motivated by a variety of interrelated goals: (a) to discuss undergraduates' performances when dealing with elementary number theory problems, with particular emphasis on impredicative problems; (b) to test the notion of semiotic control; (c) to test the APOS framework with a different population and to relate it to semiotic control; and (d) to begin an analysis of the influence of language and format in the statement of problems on students' performances.

All the data presented in this section come from Italian first-year university computer science students enrolled in an introductory modern algebra course. I have translated both the statements of tasks and their answers into English. I am aware that a thorough investigation of the multiple influences of language on students' performances should take into account the features of the specific language used in the formulation of the tasks and by students in their answers. Nevertheless, an exploratory analysis of the influence of language in mathematics learning may start from behaviors and phenomena that are common to a wide class of languages and that are not substantially altered by the translation into English. For example, confusions between the Italian words "divisibile" and "divisore" correspond well with confusions between the English words "divisible" and "divisor."

Throughout this section, the word "error" denotes students' answers regarded as inappropriate compared to the educational goals attached to that aspect of the curriculum under evaluation. It has been used for the sake of brevity and does not imply any of the pejorative attitudes often attached to it. Alternative interpretations of inappropriate behaviors are possible as well. In particular, the interpretation of "errors" as "local knowledge", that is, as the result of the application of knowledge possibly adequate to other contexts but inadequate to the actual one (see Drouhard, 1995) has proved to be a useful tool for evaluating undergraduates' behaviors. For example, most of the wrong answers to problem (1) provide good examples of local knowledge.

The Task

The following problem (problem 11) was administered to a group of thirty-nine first-year computer science students after approximately thirty hours of introductory algebra devoted to the language of sets (about four hours), the idea of function (about eight hours), introductory combinatorics (about four hours), and elementary number theory (about fourteen hours), with particular regard to divisibility, factorization, greatest common divisor, congruence, and linear Diophantine equations. There was little emphasis on various "divisibility rules." The author taught the course with an orientation toward problem solving, with the help of some senior students (tutors). Tutoring was optional. When problem solving, students were allowed to freely use books, papers, and pocket calculators. After the written test, most of the students were interviewed.

Questions (11.a), (11.b), and (11.d) are similar to some question sets proposed by Zazkis and Campbell (1996). The number M has been explicitly given (in order to avoid obstacles related to the use of variables) and its size has been chosen in order to prevent students from using a pocket calculator, or performing paper-and-pencil computations. Students are compelled to examine the number and to extract the properties they need to answer the questions. So they cannot actually perform the divisions by 63 or 18 but must use other strategies. The problems were the following.

> Let $M = 3^4 \cdot 5^3 \cdot 7^6 \cdot 10^8$. Answer to the following questions and explain your answer. (11)
> a. Is M divisible by 63?
> b. Is M divisible by 18?
> c. Is M + 5 divisible by 10?
> d. Is there an integer x such that $M \leq x \leq M + 10$ and 8 is a divisor of x?

A Priori Analysis

Students dealing with questions (11.a) and (11.b) are expected to relate the notion of divisibility to factorization. They could be more or less aware of the meanings involved. In particular, they could have learned a rule relating factorization to divisibility and answer by the simple action of factoring 63 or 18—materially searching for the corresponding factors in the given decomposition of M. This behavior, which might be regarded as an example of procedural thematization, does not imply any degree of semiotic control at all.

Question (11.c) cannot be solved by the simple inspection of the representation of M. The occurrence of the prime factors of 10 within the factorization of M + 5 cannot be empirically verified but only inferred. Students could notice that M and 5 are both divisible by 5 and thus M + 5 must be as well, that M + 5 is even (as M and 5 are odd), and finally that an even number divisible by 5 is divisible by 10. Procedural answers are not yet available, however: one could put $M = 5 \cdot K$ (for a suitable integer K), derive $M + 5 = 5(K + 1)$, and see that K + 1 must be even. In any case, this question requires the coordination of actions and processes, some of which embody knowledge and cannot be performed as actions. For example, seeing that M + 5 is even requires recalling that M is odd, that 5 is odd, that the sum of two odds is even, and coordinating these facts. From the semiotic perspective, the analysis is similar: one cannot infer M + 5 is even from its representation only—it must be interpreted using information about M.

Question (11.d) involves the recognition of the existence of an object without seeing or constructing it. Students are asked to do something like a nonconstructive proof of existence. In this specific case the construction is possible in principle, but it is much simpler just proving the existence of such an x instead of actually computing it. The mathematical contents involved in this question are not much beyond the usual primary school curricula, but students searching for a procedure generating x may get stuck. Some of the remarks on "subjective uniqueness" apply to this question as well, as students are requested to prove the existence of an object through the construction of connections involving both processes and objects.

Outcomes

The sequence of questions a, b, c, and d of problem (11) provide a reasonable ranking of students' skills, since students failing to answer a particular question in the sequence generally did not solve the following ones.

A first group of nine students did not produce any right answer. They showed very poor mastery of language and seemed unable to use words to express even the most elementary mathematical ideas and relationships. The following are some examples of wrong answers they provided for questions (11.a) and (11.b).

- "M is divisible by 63 because they have some common factor."
- "63 does not divide M because the exponents of the prime factors of 63 are not multiple of the exponents of M."
- "M is not divisible by 63 because GCD(M) $\neq$ GCD(63)."
- "M is divisible only by its prime factors and their powers."
- "M is divisible only by the product of its factors."

Answers like these show major difficulties with language: these students could not express even simple arithmetic properties and by no means were they able to link divisibility to factorization; seemingly they were trying to recall some sequence of words rather than to express some meaningful insight. All the students of this group but one did not provide any answer to (11.c). Their poor mastery of language was an obstacle to interiorization, which prevented them even from some basic physical manipulations of the symbolic expressions involved.

A second group of 14 students provided a right answer to (11.a) and (11.b) only. Some provided correct answers to (11.a) explicitly using properties of factorization; others found (by cancellation) that $\frac{M}{63} = \frac{3^4 \cdot 5^3 \cdot 7^6 \cdot 19^8}{3^2 \cdot 7} = 3^2 \cdot 5^3 \cdot 7^5 \cdot 19^8$. The use of expressions like $\frac{M}{63}$ and of procedures based on cancellations (students often used "simplify" for "divide") actually performed on paper suggests that the links they made between division and prime decomposition were procedural rather than conceptual; for some students, methods based on actual cancellations could be regarded as activities procedurally (but not conceptually) equivalent to the effective calculation of the division. Of course, activities like simplifications do not imply much semiotic control, as they do not require any interpretation of the expressions involved.

All those students providing correct answers to question (11.b) pointed out that an even number cannot divide an odd one, or that there was no factor "2" within the factorization of M. As to question (11.c), almost all the students of this group claimed M + 5 was not divisible by 10 because there was no factor "2" within M + 5, or else they tried to find out some factor "2" but performed some computations or manipulations incorrectly. This supports the conjecture that, for them, the link between divisibility and factorization hold at the level of actions only: they needed to actually see a "2" within the representation of M + 5 and actually erase it. This difficulty may be interpreted from the semiotic perspective as well: they show no awareness of the difference between a number and its representation, and since an even number must "contain" a factor "2," its representation must do the same. Question (11.c) served to break the action-based correspondence between the divisibility properties of the number and its prime decomposition, precluding the application of purely mechanical methods.

A third group of nine students gave a correct answer to (11.c) but not to (11.d). All of them gave a correct answer to (11.a) and (11.b) as well. In so doing, they used two different strategies:

(i) One student remarked that the last decimal digit of M is a "5," and then M + 5 must end with a "0," and so it is divisible by 10.

(ii) The remaining eight students explicitly wrote down expressions like $\frac{M + 5}{10}$, showed that $M + 5 = 5 \cdot (3^4 \cdot 5^2 \cdot 7^6 \cdot 19^8 + 1)$, actually cancelled the "5," and then remarked that the number within the brackets (which is the successor of an odd number) must be even.

Answer (i) is another example of thematization that takes into account properties of the representations of numbers. The decimal representation of M is not actually performed but the student can extract the piece of information he needs. All the students of this group gave wrong answers to (11.d) (whereas most of the students of the first and the second group did not answer at all). All those providing wrong answers to (11.d) claimed that there was no factor "8" in any of the numbers M, M + 1, M + 2, . . . ,M + 10. Some students added to this pattern some misinterpretations of the question: for example, "Is there an integer x such that $M \leq x \leq M + 10$ and 8 is a divisor of x?" interpreted as it were "There exists an integer x such that $M \leq x \leq M + 10$ and an integer x such that 8 is a divisor of x?"

The following excerpt from the interview of Roberto shows answers typical of a student of this group:

Interviewer: Why did you write M is divisible by 63?

Roberto: Because $63 = 7 \cdot 9$, and $9 = 3^2$, so $\frac{M}{63} = \frac{3^4 \cdot 5^3 \cdot 7^6 \cdot 19^8}{7 \cdot 3^2} = \frac{3^{\not{4}2} \cdot 5^3 \cdot 7^{\not{6}5} \cdot 19^8}{\not{7} \cdot \not{3}^{\not{2}}}$

Interviewer: And why is M not a multiple of 18?

Roberto: Because 18 is even, it means there is a factor 2, and then $\frac{M}{18}$ is a decimal number.

Interviewer: What does "$\frac{M}{18}$ is a decimal number" mean?

Roberto: I mean . . . , you know . . . , I cannot divide it, there is a remainder.

Interviewer: And why is M + 5 divisible by 10?

Roberto: Because $\frac{M+5}{10} = \frac{3^4\cdot 5^3\cdot 7^6\cdot 19^8 + 5}{10} = \frac{3^4\cdot 5^{\cancel{3}2}\cdot 7^6\cdot 19^8 + \cancel{5}}{\cancel{10}} = \frac{3^4\cdot 5^2\cdot 7^6\cdot 19^8}{2} + \frac{1}{2}$; but both the numerators of the sum, I mean $3^4\cdot 5^2\cdot 7^6\cdot 19^8$ and 1 are odd, so the sum is not a decimal number.

Interviewer: Can you explain why you answered "false" to question (11.d)?

Roberto: It is all the same, I should divide by 8, but there is no "8" I can simplify within any of these numbers.

Here, the use of fractional notation is interesting (that may result from a procedural interpretation of division oriented to the model of rational numbers) as is the conversational use of "decimal" to mean "not an integer."

Only 7 students provided a correct answer to (11.d). Five of them correctly answered all of the questions of problem 11, and 4 of them explained their answers by means of additive properties of **N** ("at least one number of the form 8 + 8 + . . . must lie between M and M + 10").

Two students came to a more complete answer by remarking that if r is the remainder of the division of M by 8, since $0 \leq r < 8$, for any value of r they could find a suitable value for x. This is a good example of thematization that coordinates both processes (the divisions—thought only but not actually performed—by 8 of the numbers from M up to M + 10) and encapsulated knowledge (the fact that $0 \leq r < 8$). One of them explicitly represented M as $8q + r$.

The remaining two students correctly answered (11.a) and (11.b) but not (11.c) because they performed some computations incorrectly. Twenty-one students did not answer question (11.d). The five students providing proper answers to all the questions (11.a–d) all used strategies depending less on actual manipulations on paper, as is shown from the following excerpt, from the interview of Alessio:

Interviewer: Why is M divisible by 63?

Alessio: Because $63 = 7\cdot 9$, and $9 = 3^2$, so all the factors of 63 are within M.

Interviewer: And why isn't M a multiple of 18?

Alessio: Because 18 is even, $18 = 2 \cdot 3^2$, and there is no factor "2" in M.

Interviewer: And why is M + 5 divisible by 10?

Alessio: Because $M + 5 = 3^4 \cdot 5^3 \cdot 7^6 \cdot 19^8 + 5 = 5 \cdot (3^4 \cdot 5^2 \cdot 7^6 \cdot 19^8 + 1)$, and $10 = 2 \cdot 5$, the factor "5" is there [*points at it*].

Interviewer: And the factor "2"?

Alessio: It is here, we said that $3^4 \cdot 5^2 \cdot 7^6 \cdot 19^8$ is odd, and so $3^4 \cdot 5^2 \cdot 7^6 \cdot 19^8 + 1$ must be even.

Interviewer: Can you explain why you answered "true" to question (11.d)?

Alessio: Because the multiples of 8 are 8 + 8, 8 + 8 + 8, . . . and at least one must fall within the interval.

Generally speaking, question (11.d) troubled students more than one would expect. The lack of a well-known algorithm appears to have induced a good number of them into providing no answer at all. Some of the errors related to language. The position of x within the second clause of the condition given in (11.d) may have deceived some of the students into misinterpreting the question. Maybe the question would have been easier had I written ". . . and x is a multiple of 8," for students' interpretation of the condition focused on x, and perhaps they expected to find it as the subject of the sentence. Some students belonging to the first two groups misinterpreted the word "divisor" and interpreted "8 is a divisor of x?" as "8 is a multiple of x?" or "$8 = x$."

DISCUSSION

These data suggest that Dubinsky's APOS scheme and Zazkis and Campbell's analysis of the development of divisibility concepts also apply to undergraduate computer science students. They also point out that different processes (like Euclidean division and simplification of prime decompositions of numbers) may be thematized as procedures but not as concepts. In other words, simplification is taken as a procedural substitute for division with no conceptual understanding. Some students have seemingly learned a strategy allowing them to simplify fractions, which means to erase a pair of matching factors, when they are on different sides of the fraction (Campbell, this volume, notes a similar phenomenon regarding "cancellation"). This seems to correspond to the need for actually performing a division rather than to the understanding of the links between divisibility and prime decomposition.

Most of the wrong answers to (11.c) and (11.d) are based on a sort of negation by failure: if one cannot apply any method he or she has learned to any of the given representations of the given numbers in order to answer the question, then he or she claims that the answer is "no." This kind of "subjective" negation seems to be a typical behavior for students who have not yet developed their divisibility concepts beyond the level of actions or even of processes. If divisibility is reduced to the actual performance of a division, or of a simplification, then the failure in doing the division must mean that divisibility does not hold. Further research is needed to test this conjecture. The use of negation by failure does not depend upon lack of experience or of logical knowledge only and occurs at any level of mathematical education including advanced ones: even mathematics majors who have passed one semester of mathematical logic keep on applying it from time to time, according to their competence in the subject they are dealing with.

On the other hand, language seems to play a very important role. The construct of semiotic control proved useful to interpret students' behaviors. In a sense, it may enrich the APOS framework, as it allows one to analyze behaviors with respect to the interpretation of statements. In particular, it provides tools to detect and analyze difficulties caused by poor mastery of language. In any case, all students who give no right answer display some trouble with language. Most of them show a very poor understanding of the basic definitions of elementary number theory and seem unable to handle even simple statements. Moreover, answers like "M + 5 is not even because there is no factor '2' " suggest that some students cannot yet distinguish between a number and its representation, and between the corresponding properties.

The simple presentation of multiple representations (discussed by Duval, 1995) with no attention to the roots of students' linguistic difficulties (in particular, those concerning natural language) is not enough to induce flexible behaviors and sometimes may only succeed in inducing students to adopt different, possibly poorer, strategies.

In conclusion, elementary number theory has proved a subject suitable for analyzing undergraduates' semiotic control of their behavior. It serves to bring forth a variety of behaviors that are not usually found with more standard abstract algebra problems. In particular, there are a fair number of students (for example, those belonging to the third group) who can produce some correct form of mathematical reasoning if they can refer to some concrete meanings, whereas they have difficulty dealing properly with more abstract subjects. Such students, who have some trouble in encapsulating some processes but have achieved an acceptable level of procedural understanding, may find elementary number theory a suitable subject to start their way into advanced mathematical thinking.

NOTES

1. According to Halliday's definition (quoted by Pimm, 1991) a register is "a set of meanings that is appropriate to a particular function of language, together with the words and structures which express these meanings."

2. Actually, Duval (1995, p. 21) uses the expression "semiotic representation register" ("registre de représentation sémiotique") as well, with the same meaning as "semiotic representation system" ("système de représentation sémiotique"). In this chapter, "register" will be used according to Halliday's definition.

3. In educational practice, sometimes different representations are shown, but in most cases only one (generally the arithmetic or algebraic code) is used to perform algorithms or solve problems.

REFERENCES

Chiappini, G., & Molinari, M. (2000). Presentazione di un metodo di approccio alle discquazioni con ragazzi di 11 12 anni. In J. P. Drouhard & M. Maurel (Eds.), *Actes des Séminaires SFIDA-9 à SFIDA-12*, (Volume III, X-45–X-54). Nice, France: IREM de Nice.

Clark, H. (1992). *Arenas of language use*. Chicago: University of Chicago Press.

Drouhard, J. P. (1995). Algèbre, calcul symbolique et didactique. In R. Noirfalise & M. J. Perrin-Glorian (Eds.), *Proceedings of VIIIème École d'Été de Didactique des Mathématiques*. Clermont-Ferrand, France: IREM de Clermont-Ferrand.

Dubinsky, E. (1991). Reflective abstraction in advanced mathematical thinking. In D. Tall (Ed.), *Advanced mathematical thinking* (pp. 95–123). Dordrecht: Kluwer.

Dubinsky, E., Leron, U., Dautermann, J., & Zazkis, R. (1994). On learning fundamental concepts of group theory. *Educational Studies in Mathematics, 27*, 267–305.

Duval, R. (1995). *Sémiosis et pensée humaine*. Bern: Peter Lang.

Ferrari, P. L. (1999). Cooperative principles and linguistic obstacles in advanced mathematics learning. In I. Schwank (Ed.), *Proceedings of the First Conference of the European Society for Research in Mathematics Education* (Vol. 2) [Online]. Available: http://www.fmd.uni-osnabrueck.de/ebooks/erme/cerme1-proceedings/cerme1-proceedings.html

Grice, H. P. (1975). Logic and conversation. In P. Cole & J. L. Morgan (Eds.), *Syntax and semantics: Speech acts* (pp. 41–58). New York: Academic Press.

Kaput, J. J. (1991). Notations and representations as mediators of constructive processes. In E. von Glasersfeld (Ed.), *Radical constructivism in mathematics education* (pp. 53–74). Dordrecht: Kluwer.

Morgan, C. (1998). *Writing mathematically: The discourse of investigation*. London: Falmer Press.

Pimm, D. (1991). Communicating mathematically. In K. Durkin & B. Shire (Eds.), *Language in mathematical education: Research and practice* (pp. 17–23). Milton Keynes, UK: Open University Press.

Vinner, S. (1997). The pseudo-conceptual and the pseudo-analytical thought processes in mathematics learning. *Educational Studies in Mathematics, 34*, 97–125.

Zan, R. (1997). Mortalita' universitaria, mortalita' matematica'. *Tracciati, 2*, [Online]. Available: http://www.eurolink.it/scuola/tracciati

Zazkis, R., & Campbell, S. (1996). Divisibility and multiplicative structure of natural numbers: Preservice teachers' understanding. *Journal for Research in Mathematics Education, 27*(5), 540–563.

6

Integrating Content and Process in Classroom Mathematics

Anne R. Teppo

Many teachers today find themselves awkwardly situated between their own past traditional school mathematics experiences and the reform-based classrooms of their students. They are being asked to enter a world far different from that they themselves experienced, where the notions are being redefined of what mathematics is and what it means to learn mathematics (Sowder, Philipp, Flores, & Schappelle, 1995). The traditional emphasis on mastering procedures related to unconnected topics is being replaced by a different perspective. School mathematics is seen as a complex, interconnected set of concepts, processes, and skills, to be learned with understanding, rather than just for mastery (Fennema & Romberg, 1999), and classrooms are envisioned as places "where all students have access to high quality, engaging mathematics instruction" (National Council of Teachers of Mathematics, 2000 p. 3).

To articulate their vision, the *Principles and Standards for School Mathematics* (NCTM, 2000) outlines a set of 10 separate content and process standards that describes "what mathematics instruction should enable students to know and do" (p. 7). Content and processes, however, do not represent "nonintersecting subjects" and the document recognizes the highly interconnected nature of the discipline, where "the areas [of content and processes] overlap and are integrated" (p. 30). Processes are learned while focusing on content and content learned within an emphasis on process.

This changing perspective to teaching with understanding and the integration of content and processes has created a need for inservice and preservice programs that provide teachers "opportunities to revisit and reconceptualize [school] mathematics" (Sowder et al., 1998, p. 107). Undergraduate programs present a first-line intervention in this area. Here, preservice students can experience first-hand what it means to engage in mathematical activity within a concept-based curriculum (Simonsen & Teppo, 1999).

This chapter describes a classroom activity based on ideas of number theory that successfully integrates content and processes in the spirit of the new *Principles and Standards*. The activity is part of a one-semester course for preservice elementary teachers designed to develop students' knowledge of mathematical content as well as provide them with standards-based learning experiences. Although many of the students in the course have had three or more years of high school mathematics, their K–12 experiences have left them with perceptions of a procedurally oriented subject that can be mastered by memorizing the appropriate collections of rules and formulas. The purpose of this chapter is to illustrate how such students can become involved in a new type of classroom mathematics.

CLASSROOM ACTIVITY

The activity provides an introduction to ideas of number theory through an investigation of patterns in natural numbers having exactly two, three, four, or five divisors. The investigation includes work with the concepts of factors, divisibility, and prime and composite numbers, as well as the processes of making and testing conjectures and expressing generalizations. The activity is also intended to help students make the transition from previous work with additive structures (addition, subtraction, multiplication, and division of whole numbers) to the multiplicative structures underlying ratio, proportion, and fractions. The activity is designed to be completed within one 50-minute class period. During that time students work individually, in small groups, and, finally, together in a whole class discussion. The instructor provides an introductory mini-lecture, coordinates student group work, and focuses the class discussion.

At the beginning of the activity, the procedures for identifying factors of given numbers and the terms "factors," "divisors," and "divides" are introduced by asking students to find the 4 different divisors for the number 15 and the 6 divisors for 45. In groups of four, students then try to find as many numbers as they can that have exactly 2 divisors, 3 divisors, 4 divisors, or 5 divisors, and to find patterns to help them add numbers to their lists.

An empty chart is drawn on the chalkboard with headings for 2, 3, 4, and 5 divisors. Students from different groups are asked to fill in particular columns in this chart with the numbers and lists of factors that their groups have found. This shared set of students' results forms the basis for a whole-class discussion that moves the lesson's focus from the calculation of particular number facts to ideas of number theory through the processes of generalizing, conjecturing, and abstracting.

Figure 6.1.
Divisor table showing student groups' initial entries.

2 Divisors			3 Divisors			4 Divisors			5 Divisors		
#	divisors	rule	#	divisors	Rule	#	divisors	Rule	#	divisors	rule
2	1, 2		4	1, 2, 4		6	1, 2, 3, 6		16	1, 2, 4, 8, 16	
3	1, 3		9	1, 3, 9		8	1, 2, 4, 8		81	1, 3, 9, 27, 81	
5	1, 5		25	1, 5, 25		10	1, 2, 5, 10				
7	1, 7					15	1, 3, 5, 15				
11	1, 11										

During group work, students do not have difficulty identifying numbers with exactly 2, 3, or 4 divisors, but many are unable to find any numbers other than 16 and 81 that have 5 divisors. Figure 6.1 is an example of the type of information provided by the student groups at this stage in the activity. Other numbers are added to the table during the class discussion as the students identify and build on patterns.

CLASSROOM VIGNETTE

The following vignette illustrates how the discussion of the divisor table promotes a range of process and content goals. The vignette is reconstructed from a detailed set of field notes summarizing an actual classroom discussion. The field notes, written by the instructor directly after the end of the class period, paraphrased the contributions of each student and the instructor but did not identify any of the speakers by name. The vignette thus recreates the essence of the classroom dialogue but is not an actual transcription of the event. While names here have been assigned to each student, they are completely arbitrary, since the sex of the original speakers was not recorded. (AT is the instructor.) The vignette provides a flavor of the pedagogical potential of the classroom activity and serves as a springboard for a discussion of the content and processes that can be drawn out of classroom work with number theory. The vignette begins after students have filled in the chart on the chalkboard, as shown in Figure 6.1.

1. AT: Does anyone see any patterns in the lists of numbers on the board?
2. Mary: The numbers with five divisors are squares of numbers with three divisors.
3. AT: [*Writes this rule on the board as it was verbally stated.*] Check this rule out with the numbers 16 and 81. Does it work?
4. Students: Yes.
5. AT: If we use the rule, can we predict another number for this column?
6. John: The number 625 is 25 squared and 25 has three divisors.

7. AT: [*Writes the five different divisors for 625 in the 5-divisor column.*] Are there any other patterns in the table?
8. Clarice: All the numbers with two divisors fit the pattern of one and the number itself.
9. Tina: For three divisors, the pattern is one, a prime number, and its square.
10. AT: [*Writes these rules out in English on the board.*] Can anyone find a pattern for numbers with exactly four divisors?
11. [*There is no response to this question.*] Tina, can you tell us what you meant by "prime number"?
12. Tina: It's any number that has exactly two divisors.
13. AT: You can see the pattern in the two-divisor column. The only factors of prime numbers are one and the number itself.
14. Sometimes it is easier to see the form of a pattern if the pattern is written symbolically instead of in words. Clarice, can you give me your rule for numbers with two divisors using symbols?
15. Clarice: It's one times x.
16. AT: [*Writes on the board "$1 \cdot x$."*] What does x represent?
17. Clarice: Any number.
18. AT: Will the number 2/5 work in your rule?
19. Clarice: It has to be any whole number.
20. AT: Tina told us that the numbers with exactly two divisors are prime numbers. Let's use the letter p to represent the number in your symbolic rule to remind us that we are using prime numbers. [*Writes on the board "$1 \cdot p$ where p = any prime number."*]
21. [*As the discussion continued, the students established that the pattern for two divisors could be represented by p, for three divisors by p^2, and for five divisors by $(p^2)^2 = p^4$.*]
22. AT: Can we identify a pattern for numbers with four divisors?
23. Janelle: The pattern goes p, p^2, then we skip a column and it's p^4. The exponent is one less than the number of divisors. The pattern for four divisors is p^3.
24. AT: [*Writes p^3 above the 4-divisor column.*] Look at the numbers and factors we've listed with four divisors. Let's test Janelle's conjecture.
25. David: 8 is 2 cubed. That fits.
26. AT: How about the numbers 6, 10, and 15? They don't fit the pattern. Was 8 just a fluke?
27. Sandra: The number 27 works. Its divisors are 1, 3, 9, and 27.
28. AT: [*Writes 2^3 next to 8 and 3^3 next to 27 in the 4-divisor column.*] Do any other numbers fit this pattern?
29. Alicia: 64 is 4 cubed. Its factors are 1, 4, 16, and 64.

30. AT: [*Writes the number 64 and its four factors in the column.*] Let's check Alicia's conjecture. Are there any other factors for 64?
31. Marsha: 64 also has factors of 2, 8, and 32.
32. AT: [*Writes out the 7 factors of 64: 1, 2, 4, 8, 16, 32, 64.*] 64 has 7 factors. Alicia, does this mean that Janelle's rule doesn't always work? Are there some cubes of prime numbers that have more than four divisors?
33. Alicia: I see what it is, 4 isn't a prime number.
34. Brenda: 64 is the cube of a perfect square.
35. AT: [*Writes* $4^3 = (2^2)^3 = 2^6$.) If you write 64 in this form it seems to fit the pattern that Janelle found earlier that the exponent is one less than the number of factors.
36. Let's go back and look at the other numbers in the four-divisor list that aren't perfect cubes. Do they fit a different pattern?
37. Kisha: The numbers can be found by multiplying two of the four factors together. For example, 6 is 2×3, 10 is 2×5, and 15 is 3×5.
38. AT: [*Circles these pairs of factors in the four-divisor column.*] Do these circled numbers exhibit a pattern?
39. Kisha: I was just looking at the list of numbers in the two-divisor column.
40. AT: Notice that the circled factors are always both prime numbers. I'd like to conjecture that the noncubed numbers in the four-factor column can be found by taking the product of two prime numbers. [*Writes the pattern "pq where q is any other prime number."*] You should be able to use this rule to find other numbers to put in the four-divisor column.

The fifty-minute class period was almost over and further discussion was postponed until the next class period.

MATHEMATICAL PROCESSES AND CONTENT

The vignette shows how the students and instructor engaged in a wide range of mathematical process, including organizing information, making generalizations from numerical patterns, making and testing conjectures, and forming abstractions. Through these processes, students worked with the concepts of factorization, divisibility, and prime and composite numbers within an underlying multiplicative structure. Verbal, written, and symbolic representations helped students to articulate and reflect on the ideas under discussion.

This section examines the nature of these processes, content, and representations. What follows is an interpretation based on the participants' observed behavior and should not be taken as an indication of actual students' cognitive constructions. Rather, the vignette is used to illuminate some of the complexity inherent in extended classroom investigation.

Organizing Information

While working in their small groups, the way that students systematized their search for and organized their lists of numbers and factors affected their ability to see patterns in the numbers. Some students were observed writing all their numbers in sequential order in horizontal lists of divisors, while other students separated numbers into groups based on the number of divisors found. As the vignette illustrates, the format of the divisor table that was written on the chalkboard made it possible to focus on patterns both within single types and among different types of numbers.

Looking for Numerical Patterns and Making Generalizations

In their groups, students quickly identified the pattern for numbers with exactly two divisors. The patterns for other numbers of divisors were not as self-evident and many students used trial and error to identify numbers for each column. In several groups, students were only able to come up with 16 as a number with exactly 5 divisors. The class discussion of the divisor table provided a structure whereby more patterns became evident.

During the discussion, several students contributed patterns for listing the divisors for all numbers within specific columns, expressing their conjectures verbally according to the characteristics of the divisors listed for each type of number. Clarice, in line 8, gave her pattern for two divisors as "one and the number itself." Tina, in line 9, expressed the rule for characterizing three divisors as "one, a prime number, and its square." In contrast, Mary (line 2) stated her conjecture in terms of the multiplicative structure of the number, rather than indicating the pattern in the listed divisors. Additionally, her rule was expressed across columns, as "the numbers with five divisors are squares of numbers with three divisors."

The process of making generalizations about the form of each rule was facilitated by recording the students' statements both in written and symbolic form. First, the rules were written on the board in the words used by each student. These sentences provided a record for future reference as well as creating a link between the students' intuitive understandings of the multiplicative structure of the different numbers and the more formal algebraic statements of the patterns' forms. Writing the patterns in symbolic form, in terms of prime number factors, made it possible for Janelle (line 23) later to generalize the multiplicative structure across columns as "the exponent is one less than the number of divisors."

Testing Conjectures

The vignette illustrates how the complexity of the patterns in the divisor table encouraged students' participation at different levels of mathematical sophistication and supported two different uses for the students' conjectures. In one use, proposed rules were tested for accuracy against specific cases in the table as well as used to generate additional numbers. For example, Mary's rule was shown to work with existing numbers in line 3 and used to find a new number in line 6. In lines 26, 37, and 40, the generality of Janelle's conjecture for *all* numbers with exactly four divisors was negated by the counterexamples listed on the board.

In a second use, the process of testing conjectures provided information about the students' understanding of both the rules and the forms of representation that were used. When the pattern for numbers with exactly two divisors was put into symbolic form, the instructor uncovered limitations with Clarice's understanding of variable. A possible interpretation of her use of x in lines 15, 17, and 19 is that she held a superficial notion of the role of a placeholder that was challenged by testing her statements using particular examples. In another situation, Alicia's test (lines 29 and 33) of the rule for finding a number with exactly four factors uncovered errors in her choice of number, rather than flaws in the stated rule.

Making and Using Abstractions

The students began to work with abstract entities rather than actual numbers when they moved from discussing multiplicative patterns for specific classes of numbers to looking for structural patterns across these classes. The rules, as students initially stated them, described properties of numbers—usually expressed in terms of particular multiplication procedures. (See, for example, Mary's rule in line 2.) Once these procedures were written symbolically, it became possible to focus the discussion on common forms across patterns. Thus, Janelle, in line 23, was able to articulate a structure shared by all the different rules.

Moving Toward Algebra

The interactions in the classroom moved from a numerical to an algebraic focus through the processes of generalization and abstraction. The students began the activity at an arithmetic level as they worked with particular number facts to construct their original divisor lists. The emphasis on finding multiplicative patterns within the divisor table then shifted the focus of attention away from executing particular calculations to describing the operations shared by many examples. The introduction of symbols encouraged a further shift to a structural level as students noted patterns common to many columns. Here, a function orientation was used to express the overall pattern for different types of number as a relation between its prime factors and the number of divisors possible for that number. Janelle's statement (line 23) epitomizes an algebraic stance where the focus of attention is on operations, with algebraic symbols being used to express relations among operations rather than to describe particular calculation procedures (cf. Esty & Teppo, 1996).

Mathematical Content

Zazkis (1999) and Zazkis and Campbell (1996a, 1996b) have found that preservice elementary teachers hold limited concept images for prime and composite numbers, the fundamental theorem of arithmetic, and notions of divisibility. They stress the importance of these interrelated topics in developing conceptual understanding of arithmetic and elementary number theory. The divisor table activity serves as a useful introduction to the multiplicative structure of the natural numbers.

One of the instructional goals of the activity was to introduce students to a new way of thinking about numbers. Up until this class period, numbers and operations had been expressed within an additive structure only (e.g., multiplication as repeated addition, and division as repeated subtraction). In this activity, the preservice teachers encountered numbers that were related multiplicatively (i.e., finding divisors, and expressing numbers as the products of prime numbers). In addition, students' number sense was extended as they characterized numbers in the divisor table as the products of prime factors. Alicia and Brenda in lines 33 and 34, for example, give examples of an extended number sense by describing 4 as "not a prime number" and 64 as "the cube of a perfect square."

Specific content encountered in the activity included the concepts of factorization, divisibility, and prime and composite numbers, and the fundamental theorem of arithmetic. During the fifty-minute class period, students worked with these concepts and used appropriate vocabulary but the concepts remained mostly implicit. The activity was intended to be an introduction, allowing students to develop richer concept images. Subsequent class periods were designed to develop explicit definitions of these mathematical entities.

Language, Communication, and Representation

It is difficult to delve deeply into the issue of language in this analysis since the vignette consists of paraphrased reconstructions of the students' utterances rather than a verbatim transcript of the discussion. However, the vignette does illustrate some of the opportunities available for mathematical communication to take place. The search for and articulation of patterns required the students and instructor to use new vocabulary such as "factor," divisor," "divides," and "prime number" in meaningful contexts. The patterns used to multiplicatively constitute different categories of numbers were articulated as students verbalized the mathematical operations involved. This language was further refined as written records of the students' verbal statements were tested with specific examples.

The vignette shows the close interrelation of process and content where, through communication, students learned both mathematics and the processes of mathematical communication, as ideas became "objects of reflection, refinement, discussion, and amendment" (NCTM, 2000, p. 60).

The activity's focus on finding patterns was also facilitated by the choices and sequencing of different forms of representation—the divisor table, written records, and algebraic symbols. At the beginning of the discussion, the structure of the divisor table made it easier to examine multiplicative structures. As patterns within columns were identified, students were able to reflect on commonalties among these patterns once they were written down. Then, the underlying mathematical structure relating the number of divisors to the number of prime factors became more evident as the rules were written algebraically. Symbolic language made it possible to stress general multiplicative relationships while at the same time ignoring particular operations such as squaring and cubing. Algebraic symbols also made it possible for the students to move to a higher level of abstraction and think about patterns of operations rather than patterns of numbers.

PEDAGOGY

This section examines how the choice of mathematical task and the promotion of reflective discourse made it possible to manage a complex set of goals within the realities of actual classroom instruction. The vignette serves as an illustration of the educational potential embedded in work with ideas of number theory.

Mathematical Task

The pattern-finding activity exhibits aspects of what the NCTM *Professional Standards* (NCTM, 1991) calls a "worthwhile mathematical task." Such a task is based on sound mathematics; engages the interest of students with a range of mathematical preparation; develops mathematical understanding, skills, and connections; and promotes mathematical communication, reasoning, and students' mathematical disposition. Added to this description is the attribute that the task allows students to quickly become engaged with a range of significant mathematical ideas and processes.

The task as it was presented to the preservice students was open-entry, in that all the students were able to participate in the initial group work, using personally meaningful methods for finding and listing factors. Since entries in the divisor table were supplied in a "bottom up" fashion from the students' own productions, discussions about elements in the table began within the students' existing knowledge bases. The subsequent discussion challenged all students with several levels of pattern recognition.

Reflective Discourse

The divisor table served as a "didactic object" that encouraged students to employ a range of mathematical processes and introduced a variety of concepts. Thompson (1998) defines a "didactic object" as something to talk about that is designed to support reflective mathematical discourse. He points out that objects are not didactic in and of themselves but are so "in the hands of someone having in mind a set of images, issues, meanings, or connections affiliated with it that need to be discussed explicitly" (p. 11), and because "of the conversations that are enabled by their presence" (p. 8).

The conversations in the vignette illustrate the notion of "reflective discourse," defined by Cobb, Bouffi, McClain, and Whitenack (1997) as being "characterized by repeated shifts such that what the students and teacher do in action subsequently becomes an explicit object of discussion" (p. 258). Two such shifts occurred in the activity. First, at the beginning of the vignette, the students described the patterns implicit in the ways they had constructed the list of column divisors. In a second shift, the discussion moved from a focus on these patterns to identifying the overall pattern across individual columns.

In line 14, the instructor implied this second shift when she said it was easier to see the "form of a pattern." However, her request to Clarice to give a rule for numbers with two divisors in symbolic form kept the discussion focused on describing the procedures for each individual rule. The difference in the subsequent discussion was that by using symbols to express the individual column rules, instead of the previously written English descriptions, the instructor was making it possible for the students to also make a shift in thinking.

The discussion moved from the activity of describing individual rules to thinking about the pattern common to all the rules in line 23 when Janelle expressed this pattern as, "The exponent is one less than the number of divisors." From that point forward, the instructor was able to keep bringing the discussion back to this level of thinking. In line 24 and lines 28 through 35 there is a "zigzag between the general and the particular" (Cobb et al., 1997, p. 273). Janelle's general rule is tested as a specific column rule and the results are then used to extend and exemplify the general rule for the case of seven divisors.

Even though the instances of reflective discourse were very short due to time limitations, the activity was carefully structured to maximize the potential for reaching this level of discussion by the instructor's progression from written sentences to symbolic notation. The vignette illustrates how this structure enabled the students to become active participants at this level. There was genuine student involvement and interaction. Although the instructor led the discussion, it did not degenerate into "a social guessing game in which the students [tried] to infer what the teacher [wanted] them to say and do" (Cobb et al., 1997, p. 269). The instructor used the students' statements to move the discussion forward.

Figure 6.2.
Extended divisor table activity.

1. Fill in the remaining rows in the Divisor Table.
   Be sure to find all the rules for each number of divisors.

2 divisors				3 divisors				4 divisors			
#	divisors	prime fac	rule	#	divisors	prime fac	rule	#	divisors	prime fac	rule
2	1, 2	$2^1 \cdot 1$	P	4	1, 2, 4	2^2	p^2	6	1, 2, 3, 6	[illegible]	[illegible]
								8	1, 2, 4, 8	2^3	P^3

5 divisors				6 divisors				7 divisors			
#	divisors	prime fac	rule	#	divisors	prime fac	rule	#	divisors	prime fac	rule
16	1,2,4,8,16	2^4	p^4								

8 divisors				9 divisors				10 divisors			
#	divisors	prime fac	rule	#	divisors	prime fac	rule	#	divisors	prime fac	rule

2. Can you find a rule for determining the number of different divisors (besides 1) for any number, given its prime factorization?

ACTIVITY EXTENSION

The vignette describes the focal point of one class period. Subsequent periods could extend this activity in different directions depending on particular instructional goals and time constraints. This section discusses one option that pursues the algebraic structure of the divisor table.

An extended divisor table (Figure 6.2) provides a didactic object for a more challenging pattern search. (Lines 24, 26, and 40 in the classroom vignette preview the direction of this activity.) First, students are asked to identify the several different rules that occur for numbers having 6, 7, 8, or 9 divisors. Once this set of patterns has been found, the challenge is to use this information to develop an algorithm for determining the number of divisors for any number, given its prime factorization. This rule requires several levels of generalization to formulate. Not only must students be able to express patterns found within and across columns in general symbolic terms, they must be able to stress and ignore certain aspects of all the different patterns to perceive a common functional relationship.

A comparison of the verbal and symbolic representations of the algorithm emphasizes the power of algebraic language (and the different functions for which variables are used) to communicate mathematical relationships.

Verbal Rule: If a number is expressed as the product of powers of its prime factors, then the number of possible divisors for that number is found by adding one to each power and taking the product of these sums.

Symbolic Rule: If a number is expressed as the product of powers of its prime factors as $N = p_1^{a_1} \cdot p_2^{a_2} \cdot \ldots \cdot p_k^{a_k}$, then the rule for the number of divisors of that number is given by $M = (a_1 + 1) \cdot (a_2 + 1) \cdot \ldots \cdot (a_k + 1)$.

Zazkis (1999) presents a derivation of this algorithm using ideas of prime decomposition and the fundamental counting principle. Her approach could generate further insights into students' work with the divisor table extension.

CONCLUSION

The ideas and processes of number theory provide effective material for promoting the type of mathematics learning advocated by NCTM's (2000) *Principles and Standards for School Mathematics*. The classroom vignette illustrates how active student participation facilitated the integration of content and process. During one fifty-minute class period, students worked with ideas from two of the Content Standards in the *Principles and Standards*—Number and Operation and Patterns, Functions, and Algebra. Enriched mathematical notions developed as students employed processes from all five of the Process Standards—Problem Solving, Reasoning and Proof, Communication, Connections, and Representation.

The divisor table promoted genuine student involvement and supported reflective discourse. The instructor played a key role in structuring this discourse by the way she organized information and recorded patterns. The pattern-finding activity brought out important ideas of numerical structure as well as modeled the processes of generalizing, symbolizing, and abstracting. This chapter demonstrates that it is possible, by thoughtful use of didactic objects such as the divisor table, to develop and manage complex learning environments in the spirit of the reform movement in school mathematics.

REFERENCES

Cobb, P., Bouffi, A., McClain, K., & Whitenack, J. (1997). Reflective discourse and collective reflection. *Journal for Research in Mathematics Education, 28*(3), 258–277.

Esty, W. W., & Teppo, A. R. (1996). Algebraic language, thinking, and word problems.

In P. C. Elliott & M. J. Kenney (Eds.), *Communication in mathematics, K–12 and beyond* (pp. 45–53). Reston, VA: National Council of Teachers of Mathematics.

Fennema, E., & Romberg, T. A. (Eds.) (1999). *Mathematics Classrooms that promote understanding*. Mahwah, NJ: Erlbaum.

National Council of Teachers of Mathematics. (1991). *Professional standards for teaching mathematics*. Reston, VA: Author.

National Council of Teachers of Mathematics. (2000). *Principles and standards for school mathematics*. Reston, VA: Author.

Simonsen, L. M., & Teppo, A. R. (1999). Using alternative algorithms with preservice teachers. *Teaching Children Mathematics, 5*(9), 516–519.

Sowder, J. T., Philipp, R. A., Armstrong, B. E., & Schappelle, B. P. (1998). *Middle-grade teachers' mathematical knowledge and its relationship to instruction.* Albany: State University of New York Press.

Sowder, J. T., Philipp, R. A., Flores, A., & Schappelle, B. P. (1995). Instructional effects of knowledge of and about mathematics: A case study. In J. T. Sowder & B. P. Schappelle (Eds.), *Providing a foundation for teaching mathematics in the middle grades* (pp. 251–274). Albany: State University of New York Press.

Thompson, P. W. (1998, June). *Didactic objects and didactic models in radical constructivism.* Paper presented at the International Conference on Symbolizing and Modeling in Mathematics Education, Utrecht, The Netherlands.

Zazkis, R. (1999). Intuitive rules in number theory: Example of "the more of A, the more of B" rule implementation. *Educational Studies in Mathematics, 40*(2), 197–209.

Zazkis, R., & Campbell, S. (1996a). Divisibility and multiplicative structure of natural numbers: Preservice teachers' understanding. *Journal for Research in Mathematics Education, 27*(5), 540–563.

Zazkis R., & Campbell, S. (1996b). Prime decomposition: Understanding uniqueness. *Journal of Mathematical Behavior, 15*(2), 207–218.

- In grades 3–5, all students should recognize equivalent representations for the same number and generate them by decomposing and composing numbers.
- In grades 6–8, all students should use factors, multiples, prime factorization, and relatively prime numbers to solve problems.
- In grades 9–12, all students should use number theory arguments to justify relations involving whole numbers.

Nonetheless, recent research (see Brown, Thomas, & Tolias, Chapter 3, this volume; Zazkis & Campbell, 1996a, 1996b; Zazkis & Gadowsky, in press) indicates that some college students, particularly prospective elementary teachers, still have difficulties recognizing and explaining divisibility relations for numbers expressed in prime-factored form. Continued research on the learning of these topics is needed, with the goal of improving teachers' abilities to guide their students in meeting the expectations above. The particular issue discussed in this chapter came to light in the process of studying how prospective elementary teachers explained the rationale behind a familiar algorithm for finding least common multiples that involves the use of prime factorizations.

In problem situations, the least common multiple of two numbers often arises intuitively as the first common entry of two ordered sequences of consecutive multiples. As a general practice, carrying out this "set intersection" method is unwieldy because it may require a large number of steps. One easy refinement is to list the consecutive multiples of the larger number while testing each new multiple for divisibility by the smaller number, stopping when a multiple of the first that is divisible by the second is found. An even more efficient and direct method is provided by comparing the prime factorizations of the two numbers. We replace the calculation of multiples with a mental comparison of the prime factorizations of the two numbers in order to find the minimal product of prime powers that contains both of their factorizations. Students often refer to this as the "higher exponent" rule.

Not surprisingly, there is a downside to using prime factorizations in this situation. While justifying that the first two methods produce the least common multiple is easy, many prospective elementary teachers (see Brown et al., this volume) cannot explain why the prime factorization method is also a valid approach. Even those who can describe when one number in prime-factored form is a multiple of another can have difficulty coordinating two such conditions to explain why the "higher exponent" rule works.

Thinking about how to help students understand the rationale behind this rule led me to formulate the problem at the beginning of this chapter. I wondered whether it might help to extend the set intersection approach and examine sequences of consecutive multiples expressed in prime-factored form. Would it be easy to recognize and reason about multiples generated in this way?

I then posed the problem in a course for prospective elementary teachers, in a course for prospective secondary mathematics teachers, in a symposium on teaching number theory for mathematicians and mathematics teacher educators,

7

Patterns of Thought and Prime Factorization

Anne Brown

A THOUGHT-PROVOKING PROBLEM

Given that there is a pattern in the sequence

$$2^2 \times 3^4, 2^3 \times 3^4, 2^2 \times 3^5, 2^4 \times 3^4, 2^2 \times 3^4, \times 5, 2^3 \times 3^5, \ldots,$$

write the next six entries in the sequence, all in prime-factored form. Having accomplished that, write the 200th term in prime-factored form, and describe a method that will provide the prime factorization of the *n*th term of the sequence.

The rest of this chapter will be more meaningful if the reader stops now and solves the problem.

THE ORIGINS OF THE PROBLEM

The strategy of examining prime factorizations to identify, compare, and contrast the multiplicative properties of natural numbers is one that students are expected to develop through their experiences in school mathematics. Indeed, the new NCTM *Principles and Standards for School Mathematics* (2000, pp. 392–393) advises instruction that enables all students to meet the following expectations.

and in informal encounters with colleagues who are mathematicians. What happened surprised me—not only did it perplex most prospective elementary teachers, almost everyone I asked to solve the problem was stymied at least briefly before recognizing the structure of the sequence fragment as that of a sequence of consecutive multiples. Though the problem is elementary, no one found it completely trivial or obvious, and its solution elicited a variety of strategies. The strategies, as well as the stumbling blocks, that I observed provide a few insights into the subtleties of the use of prime factorizations as tools for reasoning about multiplicative structure.

SOME STRATEGIES FOR SOLUTION

Perhaps the simplest solution to the problem is this: notice that the terms of the sequence fragment have a common factor of $2^2 \times 3^4$. Dividing the terms of the sequence fragment by this common factor yields the quotients 1, 2, 3, 4, 5, and 6. Since the function, $f(n) = 2^2 \times 3^4 \times n$, fits the given data, the sequence is simply the sequence of consecutive multiples of $2^2 \times 3^4$. Consequently, the nth term in prime-factored form is obtained by multiplying the prime factorization of n by $2^2 \times 3^4$.

Interestingly, this simple solution initially eluded almost all of my problem solvers; for most, it was a belated "Aha!" reaction that arose through reflection on another approach. Following are the typical solution strategies I have seen.

Strategy 1

Transform the sequence to its ordinary decimal form to obtain

324, 648, 972, 1296, 1620, 1944, . . .

from which you can see that the sequence is arithmetic with a common difference of 324. From this point, there are at least two typical paths to a solution:

- Since the first term and the common difference are both 324, it is clear that this is the sequence of consecutive multiples of 324, so the nth term is $324n$. Returning to prime factorizations, the nth term is obtained by multiplying the prime factorization of n by $2^2 \times 3^4$.
- Students who have not yet grasped the power and efficiency of prime factorizations often use the common difference to generate the subsequent six terms in decimal form: 2268, 2592, 2916, 3240, 3564, 3888, . . . and then generate a factor tree for each number. The tedium of this task inspires a search for an easier way; for some, the breakthrough comes in noticing that the 7th term is $2^2 \times 3^4 \times \underline{7}$, while for others, the realization that the terms can be generated through multiplication never comes.

Generally, more sophisticated problem solvers are not eager to transform the sequence to decimal form—they understand that the prime factorization contains useful information that will be obscured through the switch. However, some reported deciding to switch representations in order to gain information that seemed not easily observable in prime factorization form, namely, to see whether the sequence fragment is increasing. This issue of the relative size of numbers expressed in prime-factored form is also raised by some prospective elementary teachers in related number-theoretic contexts. For example, when given two numbers in prime-factored form, and asked whether one number is a multiple of another, some students are reluctant to consider the question before determining which number is larger, revealing a lack of understanding of how multiples can be recognized in this representation. This often leads to a change to decimal form, and an effort to settle the question through direct division rather than through reasoning about the information provided by the prime factorization.

In fact, for some students, prime factorizations are meaningful only as commands to perform the indicated multiplication and return to decimal form; as one of my students put it, "I need to see the actual numbers." Often, such individuals do not know how to exploit the structure implicit in prime factorizations in order to compare the terms, and they are not comfortable with performing arithmetic on prime factorizations. If that is the case, the most sensible response to this problem is to transform the sequence to decimal form. A student who solves the problem solely through the use of decimal form followed by factorization can benefit from re-examining the problem, and trying to see the solution directly from the prime factorizations.

Strategy 2

While those who used the strategy of comparing consecutive differences typically used decimal form, there is no reason that you cannot use the prime factorizations. Through this strategy, a pattern emerges:

$$2^3 \times 3^4 - 2^2 \times 3^4 = 2^2 \times 3^4\,(2 - 1) = 2^2 \times 3^4$$
$$2^2 \times 3^5 - 2^3 \times 3^4 = 2^2 \times 3^4\,(3 - 2) = 2^2 \times 3^4$$
$$2^4 \times 3^4 - 2^2 \times 3^5 = 2^2 \times 3^4\,(2^2 - 3) = 2^2 \times 3^4$$
$$2^2 \times 3^4 \times 5 - 2^4 \times 3^4 = 2^2 \times 3^4\,(5 - 2^2) = 2^2 \times 3^4$$
$$2^3 \times 3^5 - 2^2 \times 3^4 \times 5 = 2^2 \times 3^4\,(2 \times 3 - 5) = 2^2 \times 3^4$$

and the common difference of $2^2 \times 3^4$ is recognized. Very few problem solvers choose this route, perhaps suggesting that it is counterintuitive to investigate what is perceived as an additive property through the use of a representation that emphasizes multiplicative structure.

Strategy 3

Considering the possibility of a common ratio sometimes leads to an examination of consecutive ratios rather than consecutive differences. Calculating $\frac{f(n)}{f(n-1)}$ for $n \geq 2$ produces the sequence $\frac{2}{1}, \frac{3}{2}, \frac{4}{3}, \frac{5}{4}, \frac{6}{5}, \ldots$ which indicates that this is an increasing sequence defined recursively by $f(1) = 2^2 \times 3^4$ and $f(n) = \frac{n}{n-1} f(n-1)$. Most of those who looked at the sequence of consecutive ratios readily generated the next six terms of the sequence through multiplication by the appropriate ratios, but then saw the difficulty in producing the 200th term and the general term. Here is a familiar challenge that arises when reasoning inductively: how to construct a formula for $f(n)$ that is expressed only in terms of known functions of n, rather than in terms of a recursion. Of course, in this case, you can see how $f(n)$ is related to $f(1)$ by looking at the telescoping product

$$f(n) = \frac{n}{n-1} \cdot \frac{n-1}{n-2} \cdot \cdots \cdot \frac{4}{3} \cdot \frac{3}{2} \cdot \frac{2}{1} f(1) = n \cdot f(1), \text{ so that } f(n) = 2^2 \times 3^4 \times n.$$

Several of those who used the consecutive ratio approach commented that they did not immediately recognize a recursion of this form as representing the sequence of consecutive multiples of a fixed integer; perhaps its unfamiliarity should not be surprising, considering the relative simplicity of the additive version of the recursion, $f(n) = f(n-1) + 324$.

Many participants tried to identify a pattern in the exponents, with the plan of generating successive terms based on that information alone. Although people with varying backgrounds tried this, experienced problem solvers abandoned this approach much more quickly than the others. This tendency to concentrate on surface features of mathematical notation, rather than "unpacking" the notation to obtain some genuine mathematical meaning, is often seen in prospective elementary teachers (see Brown et al., this volume). In working with these students, it helps to redirect their attention to the operational underpinnings of the structure by suggesting that they consider patterns that involve arithmetic operations on the terms. Pointing out the contrast in the substance and outcome of the two strategies during a later whole class discussion helps students see how arithmetic information can be extracted from the representation of natural numbers as products of primes.

CONCLUSION

Why is this problem so puzzling, at least initially? Perhaps it is because it combines two things that we do not usually consider simultaneously: prime factorizations and the ordered sequence of all natural numbers. After all, the prime factorization of n has little necessary relation to the prime factorization of $n + 1$, beyond the fact that one contains at least one factor of 2 and the other one does not. Since the domain of a sequence is the set of natural numbers, it is possible that the sequence $1, 2, 3, 2^2, 5, 2 \times 3, 7, 2^3, 3^2, 2 \times 5, \ldots$ could play a role whenever a sequence is presented in prime-factored form. Even when it is combined with something as simple as a multiplying by a common factor, such as in $2^5, 2^6, 2^5 \times 3, 2^7, 2^5 \times 5, 2^6 \times 3, \ldots$, its tendency to mask patterns is evident.

To resolve the pedagogical issue concerning student difficulty in explaining the rationale behind the "higher exponent" rule, I decided it would be wise to rethink my approach. Recognizing that the rationale behind the method of creating successive multiples of one number while checking for divisibility by the second number is quite easy for my students, I now encourage them to extend this method to numbers represented in prime-factored form. Refined slightly, the algorithm goes as follows: to find LCM(A,B), multiply A by only those prime factors of B that are "missing" from the prime factorization of A (with their appropriate multiplicities). This process obviously creates a multiple of A because the comparison of A and B results in one or more multiplications of A by natural numbers. Since we stop once we have multiplied A by all of the prime factors of B that are missing, it is evident that the final multiple of A that is produced is the least one that is also divisible by B.

PROBLEMS FOR FURTHER THOUGHT

For the reader who is interested in additional problems similar to that presented in this chapter, here are a few possibilities:

1. Without using decimal form, find the general term for a sequence whose first six terms are:

   $2^2 \times 3^3, 2^4 \times 3^3, 2^2 \times 3^5, 2^6 \times 3^3, 2^2 \times 3^3 \times 5^2, 2^4 \times 3^5, \ldots$

2. Without using decimal form, find the general term for a sequence whose first 9 terms are:

   $2^3, 3^4, 2^4, 3^5, 2^6, 3^5 \times 5, 2^7 \times 3, 3^5 \times 5 \times 7, 2^{10} \times 3, \ldots$

3. Find the general term for a sequence whose first 16 terms are:

   1, 3, 5, 9, 7, 15, 11, 27, 25, 21, 13, 45, 17, 33, 35, 81, . . .

4. Given a sequence $\{f(n)\}$ as described in problem 3, answer the following questions:

   (a) Can you find an n such that $f(n) = 77$? $f(n) = 63$? $f(n) = 100$?

   (b) Are all of the terms of the sequence distinct?

   (c) Which natural numbers appear in this sequence?

   (d) If m is a multiple of n, is $f(m)$ a multiple of $f(n)$?

   (e) Is it true that $f(a \times b) = f(a) \times f(b)$ for all natural numbers a and b?

REFERENCES

National Council of Teachers of Mathematics (2000). *Principles and standards for school mathematics*. Reston, VA: Author.

Zazkis, R., & Campbell, S. (1996a). Divisibility and multiplicative structure of natural numbers: Preservice teacher's understanding. *Journal for Research in Mathematics Education, 27*(5), 540–563.

Zazkis, R., & Campbell, S. (1996b). Prime decomposition: Understanding uniqueness. *Journal of Mathematical Behavior, 15*(2), 207–218.

Zazkis, R., & Gadowsky, K. (in press). Attending to transparent features of opaque representations of natural numbers. In A. Cuoco (Ed.), *NCTM 2001 Yearbook: The roles of representation in school mathematics*. Reston, VA: NCTM.

8

What Do Students Do with Conjectures? Preservice Teachers' Generalizations on a Number Theory Task

Laurie D. Edwards and Rina Zazkis

Generalization has long been acknowledged as central to the practice of mathematics as a professional discipline, and has also received increasing emphasis in reform documents in mathematics education (see, e.g., National Council of Teachers of Mathematics, 1989, 1998, 2000). These documents advocate an approach to mathematics teaching in which students carry out tasks that are more closely aligned with what mathematicians actually do, as opposed to learning procedures and facts purely by rote. For example, the NCTM *Principles and Standards* state that students should be able to: "make and investigate mathematical conjectures [and] select and use various types of reasoning and methods of proof" (2000, p. 56). Furthermore, it is argued that these abilities should be developed not only in the traditional sophomore geometry course, but rather, throughout the mathematics curriculum and across the grade levels, from elementary through secondary school.

In order to support students' learning of these processes, teachers of mathematics must possess an understanding of mathematical generalization and reasoning. This understanding should not be limited to secondary school teachers responsible for formal instruction in mathematical proof; rather, all teachers of mathematics should be prepared to work with pupils as they investigate problems and reason mathematically. Thus the preparation of teachers in credential programs should, ideally, include the opportunity for preservice teachers to carry out the same processes of mathematical reasoning as those outlined in the NCTM *Principles and Standards* (2000). Mathematics content courses for preservice teachers can provide this kind of opportunity by asking students to utilize deductive and inductive arguments, to make and evaluate conjectures, and to develop generalizations based on engagement with challenging mathematical problems. The purpose of this chapter is to report on the work of a group of preservice elementary school teachers in such a course, describing their responses to an open-ended problem involving elementary number theory. In particular, we focus on how the students created and evaluated a variety of conjectures about possible general solutions to the problem. This focus, it is hoped, will extend our understanding of mathematical reasoning and conjecturing, as well as our knowledge of learning within the domain of number theory.

Although most content area studies of proof have focused on geometry, introductory number theory offers a useful and appropriate domain within which to investigate conjecturing and problem solving. One notable advantage of this domain is its accessibility: even problems with advanced solutions (for example, Fermat's last theorem) or unknown solutions (for example, Goldbach's conjecture) can be introduced and exemplified at an elementary level. More importantly, there exists a wide range of problems that can be assigned to adult learners on which progress toward solutions can be expected, if not full solutions themselves. Many such problems can be described using examples based on small numbers; such examples are easy to generate and verify and, as such, serve as important starting points in the art and science of conjecturing.

Our goal in the current study is twofold: to add to an emerging picture of the development of mathematical thinking and generalization (Chazan, 1993; Hanna, 1990; Harel & Sowder, 1996; Maher & Martino, 1996), and to document the progress and obstacles encountered by adult students as they sought a general solution to a problem involving elementary number theoretic relationships.

THEORETICAL FRAMEWORK AND RELATED RESEARCH

This work is situated within a framework that considers cognitive processes to be phenomena that emerge over time as individuals act and interact within specific social contexts (Brown, Collins, & Duguid, 1989; Lave, 1988). Mathematical reasoning, as one among many cognitive processes, is thus not characterized as an all-or-nothing capability, something that an individual either can or cannot do. Rather, we assume that this kind of thinking has emerged, culturally and individually, as a response to particular kinds of problems. Although logical and mathematical reasoning may be called upon in work or everyday contexts, for many, the development of formal mathematical thinking takes place primarily within the context of school mathematics. This context can be thought of as a subculture within a broader intellectual and social context, with norms, explicit and implicit rules, and a language all its own (Pimm, 1987; Yackel & Cobb, 1996). Mathematical learning, from this point of view, is as much a matter of acculturation, of learning appropriate modes of expression and norms for argumentation, as it is the acquisition of explicitly taught skills, results, or facts.

One goal of the research reported here is to describe the thinking of students who have been asked to work on a task that calls for mathematical reasoning, yet who are not fully enculturated into this kind of thinking. That is, the preservice teachers in this study were participants in a class in which such thinking was modeled and discussed, and in which they carried out problem solving on a variety of mathematical tasks, yet they were not "fluent" in the use of techniques for generating and testing conjectures that are standard in the discipline. They had not, in other words, fully internalized the sociomathematical norms appropriate to mathematical reasoning, and, more specifically, to working with conjectures (Harel & Sowder, 1996; Yackel & Cobb, 1996).

The development of mathematical reasoning has been explored in students of a variety of ages; however, only a few studies have investigated proof and conjecturing among preservice elementary school teachers (Martin & Harel, 1989; Pence, 1999; Simon & Blume, 1996). The current research addresses a level of reasoning that can be described as falling within the "territory before proof" (Edwards, 1997). The "territory before proof" is a metaphorical label for a "space" of potential intellectual precursors to proof: ways of thinking, talking, and acting that support the goal of seeking and establishing mathematical certainty. This space includes activities such as *noticing* and *describing* mathematical patterns or generalizations, generating *conjectures* that a given generalization is always true, *checking* a conjecture, and generating either an *inductive* or *deductive* argument for its truth. Within this broad framework, the current study focuses on a particular "neighborhood" within that territory—activities involving the generation and testing of conjectures.

The participants in the study were presented with a problem whose solution was not a specific number, but rather a rule or generalization. The problem, illustrated in Figure 8.1, was to determine the number of squares in a rectangular grid that are crossed by a diagonal for a rectangle of any integral length and width. Particular solutions to this problem can be generated by creating specific rectangles and counting; however, a full solution would consist of a formula or rule that gives the correct result for a rectangle of any length and width. The students were thus expected to move from the particular to the general in seeking a solution, an important aspect of mathematical thinking (Davis & Hersh, 1981; Mason & Pimm, 1984). As part of this process, they developed conjectures about possible rules or formulas, and tested these conjectures for generality. It is this phase of the "preproof" process that we were most interested in exploring in the research.

The general question that frames this research can be stated informally as, "What do students do with conjectures?" Given a mathematical situation in which a conjecture has been developed, one can outline, in general terms, a normative path for moving toward establishing the truth or falsity of the conjecture. One must test the conjecture, preferably using a range of examples to establish (or refute) its plausibility. Checking a conjecture for generality using a specific example results in one of two possible outcomes: the conjecture might hold true for the example, in which case, the example offers supporting or confirming evidence, or it might not. That is, the example might constitute a piece of disconfirming evidence, or what mathematicians call a counterexample. Our central research question concerns students' responses to disconfirming evidence: When an example does not support a student's conjecture, what does he or she do?

The mathematically normative response to even a single piece of disconfirming evidence is to acknowledge that the conjecture cannot be true in general. Indeed, the often-unspoken "rule" that a single counterexample is all that is necessary to refute a conjecture is one of the foundational norms of logic and mathematics. Yet this norm is not one that is taken for granted in nonmathematical, everyday reasoning, as the phrase, "the exception that proves the rule" illustrates. In many nonmathematical situations, rules are not absolute, and it is only the weight of evidence that determines whether one accepts a generalization as being true.

Thus, even when the mathematically normative response to counterexamples is modeled in a mathematics class, it may not be easily internalized by all students. Indeed, prior research has demonstrated that college students do not always treat conjecturing in a mathematically or scientifically normative way. For example, Chazan (1993) has reported on high school students' beliefs that empirical examples alone are sufficient to establish the truth of a theorem in geometry. In an experiment in which undergraduates were asked to come up with a mathematical rule relating triples of numbers, Wason stated that "there would appear to be compelling evidence to indicate that even intelligent individuals adhere to their own hypotheses with remarkable tenacity when they can produce confirming evidence for them" (1977, p. 313). These students were reluctant to discard any conjectured rule for which they were able to find confirming evidence, rather than seeking disconfirming evidence that would force them to come up with another rule.

In experiments with science students, Chinn and Brewer (1993, p. 39) identified seven responses to what they called "anomalous data." The non-normative responses were: ignoring, rejecting, or reinterpreting the data, excluding the data from the current theory, and holding it in abeyance (not rejecting it, but not using it to modify the theory either). Only two responses followed scientific or mathematical norms: first, making peripheral changes to the currently held theory, and second, making substantial changes. A similar reluctance to abandon conjectures was observed among middle and high school students who explored the composition of geometric transformations. Once they were able to find a conjecture that fit one or more of the examples they had generated, most of the students were uninterested in further testing of the conjecture, and had to be prompted to do so by the investigator (Edwards, 1997).

The current study investigates this phenomenon among adult learners within the naturalistic setting of a university course. Another feature of the study is that it utilizes the content domain of elementary number theory, and a second goal of the study is to examine the nature of the students' understanding of such number theoretic relationships as divisibility, factors, multiples, and relative primeness, relationships that arose in the course of seeking a general solution to the problem.

METHODOLOGY AND TASK

The participants in the study were undergraduates enrolled in a required thirteen-week mathematics course for prospective elementary school teachers taught by the second author. During the course, the students investigated numerical patterns in a variety of problem-solving situations (for example, Euler's formula, counting handshakes, etc.). For three-and-a-half weeks prior to receiving the assignment, the students studied topics in number theory, including factors, multiples, and prime decomposition. Although the number theoretic relationships addressed in class during this time were implicitly involved in the problem used in the study, no attention was drawn to this fact when the problem was assigned.

As a means of collecting data for the study, the students were instructed to keep a "problem-solving journal" in which they recorded all of their attempts to solve the problem. It was recommended that they work on the problem two or three times a week over a three-week period, spending thirty–sixty minutes per session. They were asked to conclude the journal with a summary of their solution attempts and a reflection on the experience. It was stated that they might not arrive at a complete solution to the problem, but that partial solutions were acceptable, and credit would be given for their efforts and reflections.

There were seventy-two students enrolled in the course, and the "diagonals" problem was one of several projects that students could choose to complete as part of the course requirements. Students were allowed to work either alone or in small groups. Twenty-three students completed individual journals, two worked as a pair, and eight worked in groups of four. A total of twenty-seven problem-solving journals were collected, representing the work of thirty-three students.

The mathematical task presented to the students is shown in Figure 8.1.

ANALYTIC SCHEME

The analysis of the students' work on this problem focused on two levels: first, the correctness of the proposed solutions and the nature of the conjectures generated by the students; and second, the students' responses to disconfirming evidence that arose during their search for a solution.

A description of a generic path through the problem, a general solution space, was created to guide our analysis of the students' solutions. This generic solution path is shown in Figure 8.2. After reading and interpreting the problem, the student must generate an initial conjecture about the relationship between the

Figure 8.1.
Generalization task.

Diagonals in a Rectangle

On squared paper draw a rectangle and draw in a diagonal.

How many grid squares are crossed by the diagonal?

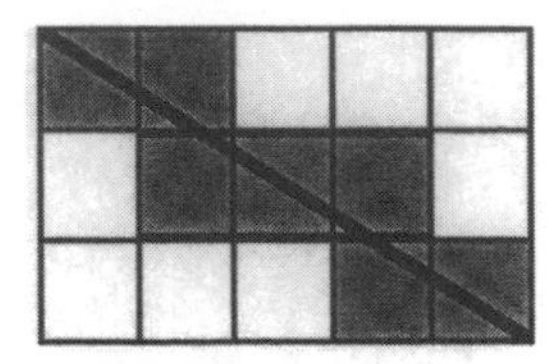

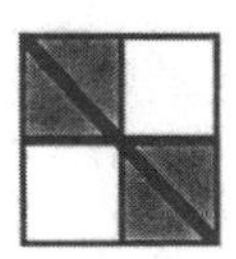

In case of a 3x5 rectangle or a 2x2 rectangle above, we can simply count.

However, can we make a decision about a 100x167 or a 3600x288 rectangle?

In general, given NxK rectangle, how many grid squares are crossed by its diagonal?

dimensions of the rectangle and the number of squares crossed by its diagonal. This conjecture is based on a number of specific examples, generated by the student in the search for a pattern or generalization. One element of the analysis considered the number of examples generated for each conjecture, and whether these seemed to be systematic or random. After stating the conjecture, the students generally went on to test it by checking additional examples. At this point, the solution space reflects two possible outcomes, one in which the example seems to support the conjecture and one in which it seems to disconfirm it. In the case of disconfirming evidence, the student might respond in one of five ways. The most acceptable or normative response, in a mathematical sense, is to reject the conjecture. Two other responses implicitly acknowledge the inadequacy of the conjecture, but do not reject the conjecture outright: holding the

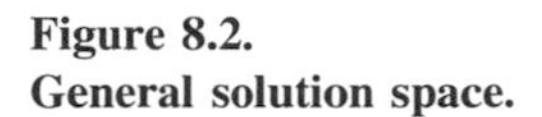

Figure 8.2.
General solution space.

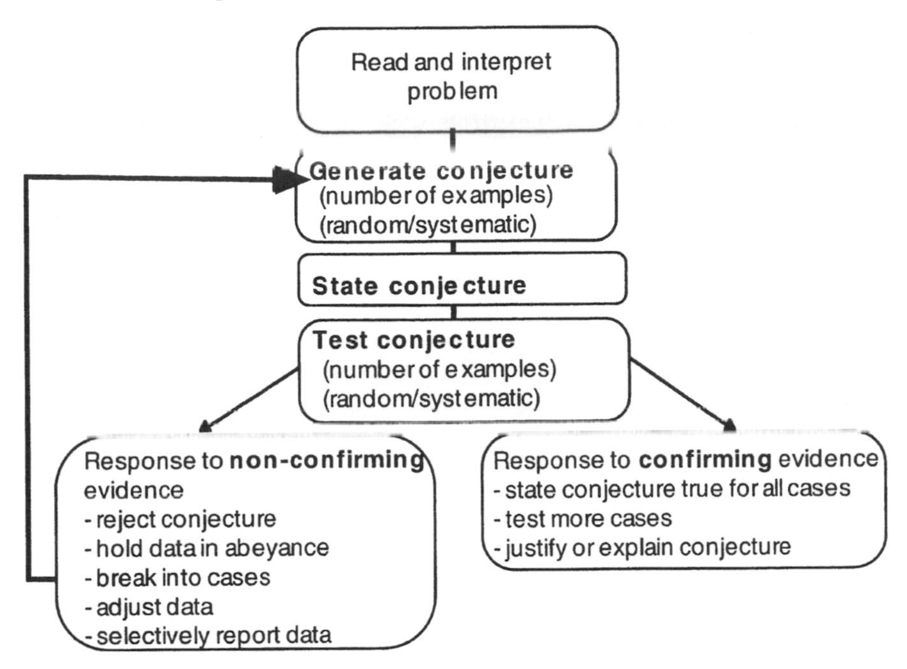

disconfirming evidence in abeyance (neither accepting or rejecting the conjecture) and breaking the conjecture into cases (that is, stating that the conjecture holds for some of the examples but not for others). Two responses are not normative, and reflect an attempt to retain the conjecture, rather than abandoning, restricting, or modifying it: adjusting the data to fit the conjecture (either intentionally or unconsciously) or selectively reporting only confirming data. Assuming the student does acknowledge that the conjecture is incorrect or incomplete, she would once again attempt to generate a new conjecture, based on additional specific examples.

The possible responses to a piece of confirming evidence include stating that the conjecture is true for all cases (quite possibly prematurely); testing more cases; and attempting to justify, explain, or prove the conjecture.

The general goal of the analysis was to explore the students' solution paths, both those that were mathematically normative and those that represented variants of a normative path, focusing specifically on students' responses to disconfirming evidence. In addition, we investigated which aspects of elementary number theory appeared in students' solutions, and the nature of their understanding of these relationships, as reflected in the context of this specific problem.

RESULTS

We first discuss the nature of the students' solutions, their problem-solving strategies, and the specific conjectures described in their journals. In seeking a solution to the problem, the students generated varying numbers of conjectures, and drew or created tables summarizing different numbers of examples. The number of conjectures described in the problem-solving journals ranged from 2 to 8, with a median of 5 different conjectures. The total number of examples varied from 4 to 44, with 25 as the median.

The students' final answers were characterized as either "complete and correct" or "partial." Complete and correct answers were general solutions that accounted for all possible rectangles, while partial solutions were conjectures that worked only in some cases (a fact that might or might not have been noted by the student). Complete and correct solutions were one of two types: the student could present a single formula or rule, or they could present a set of rules that, together, covered all cases.

The most general, single rule solution is as follows:

$$\mathbf{d} = \mathbf{n} + \mathbf{k} - \mathbf{GCD(n,k)}$$

(where **d** is the number of squares crossed by the diagonal of an $\mathbf{n} \times \mathbf{k}$ rectangle, and GCD is the greatest common divisor).

Out of the twenty-seven problem-solving journals analyzed, a total of five concluded with this single rule solution. An additional three presented complete and correct solutions, expressed as a combination of cases. The remaining nineteen journals presented partial solutions, conjectures that were true for only some cases.

A common partial solution, offered either alone or with other partial solutions, was the conjecture:

$$\mathbf{d} = \mathbf{n} + \mathbf{k} - \mathbf{1}$$

For example, given a rectangle 5 squares wide and 4 squares tall, the diagonal would cross $5 + 4 - 1 = 8$ squares. As is clear from the general solution, this conjecture is true only when N and K are relatively prime. If the examples that students generated in search of a conjecture happened to have dimensions that were relatively prime, then this conjecture would appear to be supported. This conjecture was found in eleven out of the twenty-seven journals. Another common partial solution, found in seventeen out of twenty-seven journals, was the observation that for an $\mathbf{n} \times \mathbf{n}$ rectangle (i.e., a square), the diagonal crosses **n** squares.

Figure 8.3.
Visual decomposition strategy.

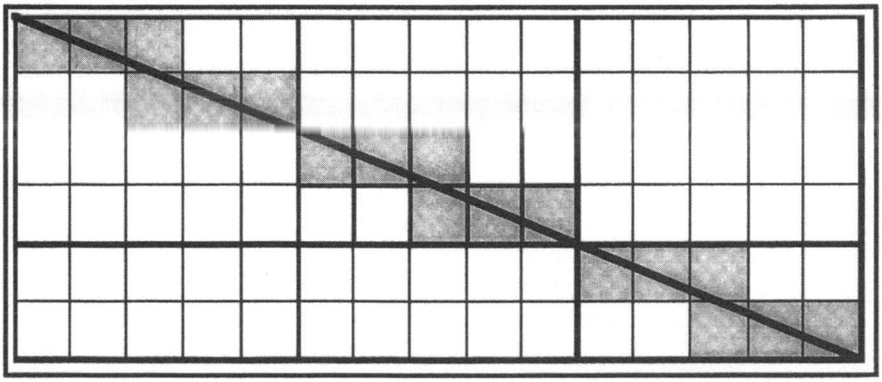

The "Visual Decomposition" Strategy

When the sides of a given rectangle are not relatively prime, then the **n** + **k** − **1** conjecture fails. In six of the journals, after the students noted this fact, they went on to analyze the visual composition of rectangles whose sides are not relatively prime, that is, cases in which **n** and **k** have a common factor.

The "visual decomposition strategy" and its relationship to the greatest common divisor of the two sides are illustrated in Figure 8.3. In this example, a 6 × 15 rectangle is decomposed into similar, smaller rectangles, by dividing the length and width of the original by 3 (the greatest common divisor of 6 and 15). This results in 9 smaller rectangles, each measuring 2 × 5. The diagonal of the original rectangle passes through three of these rectangles. Thus if the number of squares crossed by a diagonal of the 2 × 5 rectangle is 2 + 5 − 1 = 6, then the number of squares crossed in a 6 × 15 rectangle is 6 × 3 = 18.

In general, such a partition is always possible when **n** and **k** have a common factor. Each small rectangle will be similar to the **n** × **k** rectangle, scaled by the **GCD(n,k)**. The dimensions of each small rectangle will thus be **n/GCD(n,k)** × **k/GCD(n,k)**. Since these dimensions are derived by dividing the length and width of the original rectangle by their greatest common divisor, the dimensions of the small rectangles are relatively prime. Thus, the number of squares crossed by the diagonal of a small rectangle would equal **n/GCD(n,k)** + **k/GCD(n,k)** − **1**. The diagonal of the original rectangle passes through **GCD(n,k)** of the smaller rectangles; thus, the number of squares crossed by its diagonal is given by:

$$\mathbf{d = GCD(n,k) \times (n/GCD(n,k) + k/GCD(n,k) - 1) = n + k - GCD(n,k)}$$

The same set of observations could lead to the following set of rules:

1. When **n** and **k** are relatively prime, then $\mathbf{d} = \mathbf{n} + \mathbf{k} - \mathbf{1}$
2. When **n** and **k** have a common factor **F**, then $\mathbf{d} = \mathbf{F} \times (\mathbf{n/F} + \mathbf{k/F} - \mathbf{1})$

In fact, rule (1) represents a special case of rule (2) because when **n** and **k** are relatively prime then

$$\mathbf{F} = \mathbf{GCD(n,k)} = \mathbf{1}.$$

The six journals written by students who actually arrived at this strategy did not take a formal approach like that outlined above. Instead, the students noted visual patterns within the particular examples they generated. In particular, the students noted the existence of smaller rectangles formed when the large rectangle's diagonal crossed precisely at the corner of an interior square, or simply focused on the number of places where such an intersection occurred. One student only discovered the relevance of the greatest common divisor after spending five days on the problem with no real progress. She then "went back and made my drawing pretty" by "doodling" and "adding to them [*sic*]." After this doodling, she noticed the visual decomposition into similar rectangles, and asked herself about "the significance of say a 1×2 inside a 4×8 a total of four times." After creating a number of similar examples, she generated the GCD conjecture. In a total of three cases, students were able to use similar visual reasoning to discover the most general, or "GCD," solution described above.

Numerical Pattern-Seeking Strategies

The visual decomposition strategy appeared in a minority of problem-solving journals (six out of twenty-seven). For the most part, the students attempted to generate conjectures not through close examination of the visual configurations found in the rectangles, but by looking for a numerical relationship among some combination of features in the diagrams. These features included length and width of the sides, total number of squares (area) of the rectangle, the ratio between squares crossed by the diagonal and squares not crossed, and the length of diagonal. A common strategy was to consider separate cases for when the side lengths were even and odd. This strategy, pursued in seven journals, resulted in the following set of rules:

(1 e/o) If both **n** and **k** are even, then $\mathbf{d} = \mathbf{n} + \mathbf{k} - \mathbf{2}$;
(2 e/o) otherwise (that is, if both odd or at least one odd), then $\mathbf{d} = \mathbf{n} + \mathbf{k} - \mathbf{1}$.

For some students this pattern served as an intermediate conjecture; for others it constituted a final solution. If the students chose pairs of even numbers for which 2 is the GCD, rule (1) would appear to be confirmed, while pairs of

Table 8.1.
Summary of Solutions.

"Complete and correct" solutions	Single formula: 5 By cases: 3
solutions with $n \times n$	17
solutions with $n + k - 1$	11
solutions with visual GCD decomposition	leading to complete solution: 3 no complete solution: 3
solutions mentioning GCD, no visual link	5
solutions mentioning even/odd	7
solutions mentioning factors/multiples (other than GCD)	3

relatively prime numbers would seem to confirm rule (2). Given the high proportion of confirming evidence for (1) and (2) with numbers chosen at random, the special features of these pairs were not readily recognized.

In five journals, students arrived at the greatest common divisor solution purely by examining numerical relationships in the examples they drew, without explicitly mentioning similar rectangles, or other visual features. An additional three journals mentioned factors or multiples (other than the GCD or even and odd numbers) as being involved in the solution.

Table 8.1 summarizes the solutions involving each type of conjecture found in the twenty-seven journals.

Responses to Disconfirming Evidence

We turn now to an analysis of students' responses when they generated an example that did not support their current conjecture. As noted above, a mathematically normative, or logically correct, response to an example that contradicts a conjecture would be to reject the conjecture, or to restrict it to the given cases for which it is true. It is also possible to hold the data in abeyance; using it neither to confirm or disconfirm the current conjecture, but continuing to test examples and search for a general solution. Invalid responses to disconfirming evidence would include adjusting data to fit the conjecture (either consciously or unconsciously), or selectively reporting only data that is consistent with the conjecture. The latter two responses may reflect reluctance on the part of students to relinquish a conjecture they have worked hard to find, or a desire to bring closure to a problem they find difficult to solve.

All five responses described above were found among the students who worked on this problem. Student DK, for example, represents a normative response. She started by drawing 16 examples, the first few random, but the rest chosen in an increasingly systematic fashion. After stating that she saw no pattern in these examples, she tried using the Pythagorean theorem to calculate **d**, but discarded this approach after trying two examples. She then created a table showing eight drawings of rectangles with dimensions ranging from 2 × 3 to 2 × 10, and noted that in each case, **d** equaled either **n** + **k** − **1** or **n** + **k** − **2**. She called the term subtracted from **n** + **k** in her table **X**, and asked herself, "What is X?" She then drew another table with seven systematic examples, and conjectured that **X** was the GCD of **n** and **k**. She tested this with three examples (two with common factors and one whose sides were relatively prime), and concluded that the conjecture was true.

Student JJ presents a contrasting example. JJ initially created a large and well-organized table summarizing forty-one drawings (not all of which were shown in the problem-solving journal), and generated the conjecture that **d** = **n** + **k** − **1** (making an exception for **n** × **n** squares). She then tested this conjecture for three cases: where **n** is even and **k** is odd, both are odd, and both are even. Discovering that her formula seemed to fail only when both dimensions were even, she re-examined these cases, and generated a second conjecture, that **d** = **n** + **k** − **2**. She tested this with three examples (12 × 19, 6 × 10, 5 × 15) and found that it worked only for the even–even case. She also "discovered" that her first conjecture, **n** + **k** − **1** worked for the other two cases. Unfortunately, this discovery was based on a miscoloring of the grid squares in the third example in a way that appeared to confirm the conjecture. There is no way to know whether this miscoloring was intentional or not; the squares that were miscolored were very close to the diagonal, and, given the scale of the grid paper used by JJ, she may have made a legitimate error. Believing that she had covered all cases, JJ presented three formulas in her summary: one for **n** × **n** squares; one for odd–odd and odd–even side lengths, and one for even–even side lengths. She gave no evidence of awareness that this was not a complete solution, nor did she seem aware of the two mistakes in her original table of examples that happened to support her conjectures.

One aspect of students' work with conjecturing and problem solving is their awareness of whether their solutions are complete; that is, whether further testing might reveal counterexamples to their final conjecture. Among the nineteen journals containing only partial solutions, nine journals contained explicit statements acknowledging that the solutions might not be complete. In the other ten journals, there was no indication that the students were aware that their solutions were incomplete. Among the eight journals containing correct solutions, only one student stated explicitly, "This is not a proof. There could be some scenario I haven't thought of yet," when presenting her final conjecture. A second student, while not making a similar explicit statement, did present an argument (which she labeled "Logic") in which she attempts to explain why her solution worked. The remaining six journals presented the correct solution with no discussion of whether further testing or a proof might be needed.

Table 8.2.
Responses to First Occurrence of Disconfirming Evidence.

Reject current conjecture	12
Restrict or break into cases	10
Ignore evidence	1
Hold in abeyance	2
Report data selectively	2
Adjust data (*at a later point in the journal)	5*

In terms of the focus of the research, we were particularly interested in the students' response when a given example contradicted his or her current conjecture. Although some of the conjectures generated by the students were phrased in vague, untestable terms (e.g., "The ratio of length and width may have something to do with it"), most were specific testable formulas or verbal rules. To analyze responses to disconfirming evidence, we examined the first point in each journal at which the student faced an apparent counterexample to their current conjecture. In twelve out of twenty-seven cases, the students rejected the conjecture, while in ten journals, they restricted the conjecture or broke the problem into cases.

In addition to these mathematically acceptable responses, one student ignored the disconfirming evidence, claiming that it fit his formula when his drawing did not in fact support this claim. Two students also seemed to ignore or hold the disconfirming evidence in abeyance, continuing to use the same conjecture to test additional examples. Two students seemed to have reported the data selectively, since no examples contradicting their (incorrect) conjectures were presented. Another possibility is that these students simply did not test enough examples, although since their conjectures were presented with only a small number of (confirming) examples, it seems likely that they wanted to present a "correct" solution, and were willing to "edit" the data to do so.

In addition to students selectively reporting only data that confirmed their current conjecture, we found five journals in which the data appeared to be adjusted to fit the conjecture, although this did not necessarily occur at the first encounter with disconfirming evidence. As with student JJ, this "data adjustment" was done primarily by filling in and counting grid squares that were not actually crossed by the diagonal, but which were "close." Students' responses to the first occurence of disconfirming evidence are summarized in Table 8.2.

Specific Influences on Students' Behavior

The students who participated in this study had little prior experience working with elementary number theory, or with conjecturing and proof. They tended to take an approach characterized by "naive empiricism"—seeking a formula that would tie together the collection of specific examples they generated, rather than working from established results or an analysis of the structure of the problem. Thus, they tended to focus on "crunching numbers" and not on deeper aspects of the problem. This focus was not limited to the unsuccessful students—even some students who arrived at the general "GCD" solution were not satisfied with it. They sought a formula containing only **n** and **k** as variables, possibly viewing "**GCD(n, k)**" as an instruction to compute something rather than as a valid element of an algebraic expression.

References to even and odd numbers appeared frequently in students' journals. By recognizing that the relationship between the chosen numbers for **n** and **k** somehow played a role in determining **d**, students' immediate attention was turned to the evenness or oddness of the side lengths. However, despite recognizing that this separation into cases was not comprehensive, and did not lead to a solution, the students had difficulty generalizing the idea of "evenness" to the idea of "divisibility" by other numbers, that is, to the factors of **n** and **k**. For example, one student, considering the 4×8 rectangle, noticed that one of her rules for even/odd numbers ("if both **n** and **k** are even, then $\mathbf{d} = \mathbf{n} + \mathbf{k} - \mathbf{2}$") did not always work. With this example in mind, she generated another rule: if **n** is a multiple of **k**, then $\mathbf{d} = \mathbf{n} + \mathbf{k} - \mathbf{k}$. She confirmed this conjecture, referring to it as a special case. However, she didn't proceed to considering other factors of the side lengths, or common factors.

Zazkis (1998) observed a similar resistance or inability to generalize the idea of evenness to the idea of divisibility among a similar population of students exploring other problems in number theory. A possible source for this difficulty is that students do not think of the property of being an "even" number as equivalent to being "divisible by 2." Rather, Zazkis's work has shown that students often think of evenness as a property of the last digit of a number, and take note only of whether a number is odd or even, as if these two possibilities exhaust the different "types" of possible whole numbers. It has been suggested that emphasizing the equivalence between evenness and divisibility by 2 could help in generalizing from divisibility by 2 to divisibility by other numbers. Along the same lines, we suggest here that pointing out that the property "both **n** and **k** are even" is equivalent to the property "**n** and **k** have a common factor of 2" could help students in attending to common factors of other pairs of numbers.

DISCUSSION

Examining the students' work on this challenging problem as a whole, we find that complete and correct solutions were presented in only eight out of the twenty-seven journals. On the other hand, if we revisit our main research question and ask, "What do students do with disconfirming evidence?" the picture is somewhat more encouraging. The vast majority of students (twenty-two out of twenty-seven journals) responded appropriately the first time they encountered evidence that did not confirm their conjecture. That is, they either rejected the conjecture completely, or restricted it and sought a new conjecture that worked in other cases. Five students made statements that suggested they were very reluctant to give up their conjectures, and in fact either ignored or, possibly, selectively presented data in order to provide more support for their formulas.

The fact that these students were enrolled in a course in which normative mathematical reasoning was both modeled and explicitly discussed may help to account for the fact that most were willing to reject a conjecture once a counterexample surfaced. In fact, the instructor (the second author) stressed to the students that "true in 99 percent of cases" means "false in mathematics." The students had clearly heard this; in their journals, students made comments like, "Since we have one false statement, it's not going to work." However, the fact that some students ignored this rule, and others seemed to need to remind themselves of it explicitly, suggests that although they could remember and recite the idea, only a few had internalized it. Nearly one third of the journals (eight out of twenty-seven) showed evidence of what might be called "wishful thinking" on the part of students who, consciously or unconsciously, buttressed their conjectures in ways that were mathematically invalid, by either ignoring, adjusting, or selectively reporting data that did not support their current conjectures.

One of the difficulties in solving this kind of challenging problem is knowing which features of the situation are relevant and which are not. For the most part, the students attacked the problem in a number-driven fashion, rather than by analyzing visual or structural features, which might help them. Very few noticed the smaller similar rectangles nested within a rectangle whose sides had common factors. In fact, few students noted factors or multiples as a salient aspect in their solutions. Many more took notice of whether the sides were even or odd numbers. As noted above, this "even/odd" distinction seems to be quite robust, and students had difficulty in generalizing from even/odd cases to examine divisibility by other integers, a move that would have been effective in the current problem.

CONCLUSION

The findings of this study shed light on the range of students' responses to disconfirming evidence when exploring conjectures related to a challenging mathematical problem whose solution involves elementary number theory. There are obvious limitations to the use of written journals as a data source, since much must be inferred about students' thinking from what they choose to present on paper. In future research, it would be useful to follow up on the analysis of students' journals with selected individual interviews as a means of clarifying students' intentions and thinking during the course of their work on this problem. In addition, in a clinical interview setting, it might be possible to investigate how far a student is able to go with a problem like this; that is, to examine thinking that goes beyond the verification of conjectures to the construction of explanations and justifications.

REFERENCES

Brown, J. S., Collins, A., & Duguid, P. (1989). Situated cognition and the culture of learning. *Educational Researcher, 18*(1), 32–42.

Chazan, D. (1993). High school geometry students' justification for their views of empirical evidence and mathematical proof. *Educational Studies in Mathematics, 24*, 359–387.

Chinn, C. A., & Brewer, W. F. (1993). The role of anomalous data in knowledge acquisition: A theoretical framework and implications for science instruction. *Review of Educational Research, 63*(1), 1–49.

Davis, P. J., & Hersh, R. (1981). *The mathematical experience*. Boston: Houghton Mifflin.

Edwards, L. D. (1997). Exploring the territory before proof: Students' generalizations in a computer microworld for transformation geometry. *International Journal of Computers for Mathematical Learning, 2*, 187–215.

Hanna, G. (1990). Some pedagogical aspects of proof. *Interchange, 21*(1), 6–13.

Harel, G., & Sowder, L. (1996). Classifying processes of proving. In L. Puig & A. Gutierrez (Eds.), *Proceedings of the Twentieth International Conference for the Psychology of Mathematics Education* (Vol. 3, pp. 59–66). Valencia, Spain: Universitat de València.

Lave, J. (1988). *Cognition in practice*. Cambridge: Cambridge University Press.

Maher, C., & Martino, A. (1996). The development of the idea of mathematical proof: A 5-year case study. *Journal for Research in Mathematics Education, 27*(2), 194–214.

Martin, W. G., & Harel, G. (1989). Proof frames of preservice elementary teachers. *Journal for Research in Mathematics Education, 20*(1), 41–51.

Mason, J. & Pimm, D. (1984). Generic examples: Seeing the general in the particular. *Educational Studies in Mathematics Education, 15*, 277–289.

National Council of Teachers of Mathematics. (1989). *Curriculum and evaluation standards for school mathematics*. Reston, VA: Author.

National Council of Teachers of Mathematics. (1998, October). *Principles and standards for school mathematics: Discussion draft*. Reston, VA: Author.

National Council of Teachers of Mathematics. (2000). *Principles and standards for school mathematics*. Reston, VA: Author.

Pence, B. (1999). Proof schemes developed by prospective elementary school teachers enrolled in intuitive geometry. In F. Hitt & M. Santos (Eds.), *Proceedings of the Twenty-first Annual Meeting of the North American Chapter of the International Group for the Psychology of Mathematics Education* (Vol. 2, pp. 429–435). Columbus, OH: ERIC Clearinghouse for Science, Mathematics, and Environmental Education.

Pimm, D. (1987). *Speaking mathematically: Communication in mathematics classrooms.* London: Routledge & Kegan Paul.

Simon, M., & Blume, G. (1996). Justification in the mathematics classroom: A study of prospective elementary teachers. *Journal of Mathematical Behavior, 15*(1), 3–31.

Wason, P. C. (1977). "On the failure to eliminate hypotheses . . ."—a second look. In P. N. Johnson-Laird & P. C. Wason (Eds), *Thinking: Readings in cognitive science* (pp. 307–314). Cambridge: Cambridge University Press.

Yackel, E., & Cobb, P. (1996). Sociomathematical norms, argumentation, and autonomy in mathematics. *Journal for Research in Mathematics Education, 27*(4), 458–477.

Zazkis, R. (1998). Odds and ends of odds and evens: An inquiry into students' understanding of even and odd numbers. *Educational Studies in Mathematics, 36*(1), 73–89.

9

Generic Proofs in Number Theory

Tim Rowland

My purpose in writing this chapter is to draw attention to the relative ease with which the domain of number theory lends itself to generic argument, presented with the intention of conveying the force and the structure of a conventional generalized argument through the medium of a particular case. I shall also contend that deliberate deployment of particular-but-generic strategies should be more commonplace in classrooms and in textbooks, in order to convince students of the truth of number-theoretic theorems and student-generated conjectures. In effect, I am saying that the potential of the generic example as a didactic tool is virtually unrecognized and unexploited in the teaching of number theory, and I am urging a change in this state of affairs.

I offer some examples of what generic exposition and generic proof might look like. Clearly these will be subject to the limitations of the written word and will be more like a textbook account than an interactive exposition. The examples of generic proof in this chapter are initially drawn from or inspired by the literature—mathematics, mathematics education, and popular exposition of mathematics. I shall proceed with further examples from my own teaching, in which I deliberately set out to implement generic approaches in the context of a university undergraduate class, before proposing a set of guiding principles underpinning the construction and presentation of such generic proofs in number theory. The chapter concludes with consideration of generic proofs as a pedagogic strategy in the teaching of number theory.

PURPOSES OF PROOF

While I would readily agree that proof is a *sine qua non* of mathematics, it is salutary to bring to mind the plight of the legions of students who never appropriate the procedures of formal mathematical proof for themselves. Those who commit the proofs of standard theorems to memory in order (hopefully) to reproduce them in exams but who are unable to see the architecture of the arguments. Those who despair of constructing their own proofs, who dread and avoid exam questions that include the injunction "Prove that. . . ." Logico-deductive proof is, after all, a cultural artifact and is often a barrier between those able to use it and those who struggle to do so.

As a rule, students are expected to acquire knowledge of mathematical proof procedures by a process of osmosis (one *could* dignify the process by speaking of "apprenticeship"). In a sense, this is not unreasonable. While it is possible to classify and "teach" a range of possible proof-techniques—contradiction, exhaustion, mathematical induction, and so on—simply knowing that these possibilities exist just doesn't seem to help with the active construction of proof. Worse, in the United Kingdom, proof (as a topic) is commonly encountered as proof by mathematical induction (PMI) by high school students, usually in connection with summation of finite series. It is not insignificant, I believe, that PMI can be reduced to a kind of proof-algorithm. Check $n = 1$, assume the proposition is true for n, and so on. As such, it is a grotesque parody of the enterprise of mathematical proof in general, for proof, far from being algorithmic, is a form of problem solving *par excellence*. On entry to university, such students then seem to expect that every proof can be accomplished using PMI. At the very least, they believe that a proper proof must be dense with algebraic symbolism and must avoid, if possible, sentences in any natural language. For all but the most successful students, proof remains inaccessible as a dimension of their active mathematical lives. It belongs to "other people's mathematics," textbook mathematics, lecturers' mathematics, with which they have, at best, a passive, submissive relationship. Thus, proof serves typically not to empower but to disenfranchise, to alienate.

Do we shrug our shoulders and complain that mathematics students aren't what they used to be, or shall we consider whether the "problem" is ours as well as the students'? If we have a part-share in the problem, what can we do about it? First, it is important to be clear about why proof is important in mathematics, and to be clear that there is more than one such "why." Sometimes the form of a proof needs to be tailored to meet the particular reason for proof that we are addressing on any occasion. Indeed, the *purposes* of proof need careful consideration in order to achieve some consensus on what might be acceptable (let alone appropriate) as a proof in a given context.

Reuben Hersh (1993) argues that the role of proof in the classroom is different from that in research; Hanna (1989) makes the same point. For research mathematicians, claims Hersh, the purpose of proof is *conviction*, by which he means that the formal proof is the guarantor of the "truth" of a mathematical proposition.[1] Support for such a view is provided in a recent best-selling book (Singh, 1997) celebrating Andrew Wiles's recent[2] proof of Fermat's Last Theorem. In the practice of mathematical research, as Hersh points out, mathematical proofs are submitted to the scrutiny of qualified (human) judges. It is an interesting case of peer review within a particular community of practice. Approval by expert gatekeepers amounts to the achievement of conviction for the community as a whole, most members of which are unlikely ever to read the proof for themselves. The effectiveness and rigor of the process was highlighted by the progress of Wiles's 200-page proof once it had been submitted to *Inventiones Mathematicae* for publication. One of the referees detected a fundamental error in the argument, and Wiles needed another year to fix it (Singh, 1997).

The situation in the classroom, says Hersh, is very different. "In the classroom, convincing is no problem. Students are all too easily convinced. Two special cases will do it" (1993, p. 396). Hersh argues that in the teaching context, the primary purpose of proof is to *explain*, to illuminate why something is the case rather than to be assured that it is the case. (The Pythagorean proposition is a cornerstone of mathematics, and no student should doubt its truth; but a proof[3] serves to shed light on its inevitability.)

By way of an example of classroom practice, consider one rich mathematics task, adapted from Foxman, Ruddock, Badger, and Martini (1982, pp. 102–111).

> **Partitions**. The number 3 can be "partitioned" into an ordered sum of (one or more) positive numbers in the following four ways: 3, 2 + 1, 1 + 2, 1 + 1 + 1. Find all such ordered partitions of 4. In how many ways can other positive numbers be partitioned?

The mathematics teacher introduces this task to the class, and organizes discussion in pairs. Work on this activity soon produces some data: as well as the 4 given partitions of 3, they find that there are 2 partitions of 2, and 8 possible partitions of 4. Emma notices that as the number to be partitioned increases from 2 to 3 to 4, so the number of partitions doubles from 2 to 4 to 8. Her prediction that there will be 16 partitions of 5 is subjected to empirical confirmation. Cathy goes on to make the conjecture that "this always happens." The conjecture is arrived at by process of *inductive* inference.

The teacher calls the class together for plenary discussion of the problem. There is consensus about the universal validity of the doubling pattern. The epistemological issue at this point is not one of conviction but of insight. *Why* is it that the number of partitions doubles at each stage? The teacher develops an explanation of the doubling—a *deductive* argument such as the following. Consider any partition of n. If I increase by 1 the size of the last part, I have produced a partition of $n + 1$. If instead, I adjoin an additional part of size 1, I have produced a different partition of $n + 1$. So there are at least twice as many partitions of $n + 1$ as there are of n. Further argument establishes that this process accounts for every partition of $n + 1$. This constructive proof explains the observed doubling phenomenon.

GENERIC EXAMPLES

The argument above (relating partitions of $n + 1$ back to those of n) can be effectively presented, from the point of view of both explanation *and* conviction, by assigning a particular value to n, say 3. The exposition then describes how each actual partition of 3 begets two identifiable partitions of 4. Indeed, my experience with students indicates that careful scrutiny and comparison of the 4 partitions of 3 alongside the 8 partitions of 4 can trigger explanatory insight concerning the way that each partition of 3 is related to 2 partitions of 4. Such an argument amounts to proof by generic example (Balacheff, 1988; Mason & Pimm, 1984).

> The generic proof, although given in terms of a particular number, nowhere relies on any specific properties of that number. (Mason & Pimm, 1984, p. 284)
> A generic example is an actual example, but one *presented in such a way* as to bring out its intended role as the carrier of the general. (emphasis added, Mason & Pimm, 1984, p. 287)

A video[4] published by the British *Office for Standards in Education*, intended to exemplify and promote "direct teaching" of mathematics in schools, features one teacher, Kate, with a class of ten- and eleven-year-olds. In the middle phase of the "Numeracy Hour" lesson, Kate introduces the "Jailer Problem."

> A certain prison has 100 prisoners, 100 cells, and 100 jailers. One prisoner is assigned to each cell. One night, when the prisoners are all locked away, the first jailer unlocks all the cells.[5] Then the second jailer locks all the cells whose numbers are multiples of 2. Next, the third jailer changes the state of all the cells that are multiples of 3 (so that 3 is locked, 6 unlocked, and so on). The fourth jailer then changes the state of all the cells that are multiples of 4, and so on through to the 100th jailer. The jailers then all fall asleep. Which prisoners were able to escape from their cells?

The children work on the problem in small groups, before being brought together by Kate to share and review their findings. It soon becomes evident that the solution hinges on the fact that every square number has an odd number of factors. In fact, Kate explains this to the class by reference to (what we recognize as) a generic example. She points out that every factor of 36 has a distinct co-factor, with the exception of 6, and so it must follow that 36 has an odd number of factors. She then generalizes, "One of the factors of a square number is a number times itself [sic]; that's why it's a square number, isn't it?" No reference is made in the commentary to this aspect of her teaching and proof strategy. As Balacheff (1988) so clearly and elegantly puts it: "The generic example involves making explicit the reasons for the truth of an assertion by means of operations or transformations on an object that is not there in its own right, but as a characteristic representative of the class" (p. 219).

By way of contrast, consider the (false) proposition that $n^2 + n + 41$ is prime for all natural numbers n. I may confirm the truth of the instance when, for example $n = 30$, by evaluating $30^2 + 30 + 41$, which is 971, and checking that no prime from 3 to 31 divides 971. But this gives no insight whatsoever as to why $n^2 + n + 41$ might be prime for any other value of n.

A generic example serves not only to present a confirming instance of a proposition—which it certainly is—but also to provide insight as to *why* the proposition holds true for that single instance. A transparent presentation of the example is such that analogy with other instances is readily achieved, and their truth is thereby made manifest. Ultimately the audience can conceive of no possible instance in which the analogy could not be achieved. A fuller account of the psychological and epistemological status of generic examples is given in Rowland (1999).

EXAMPLES FROM THE LITERATURE

Closely related if not identical to proof by generic example are the notions of "action proof" (Morley, 1967; Semadeni, 1984) and "illuminating example" (Walther, 1984, 1986). It is interesting to observe how these and other writers have drawn on number-theoretic content in order to illustrate how generic examples might point to general arguments. I offer a few cases in point.

Summation of Consecutive Odd Numbers

Semadeni (1984) credits Arthur Morley with coining the name "action proof" for an argument involving actions,[6] which, while performed on particular objects, can be "seen" to be extendable beyond the case(s) offered for inspection. Morley's example is the summation of consecutive odd numbers beginning with 1.

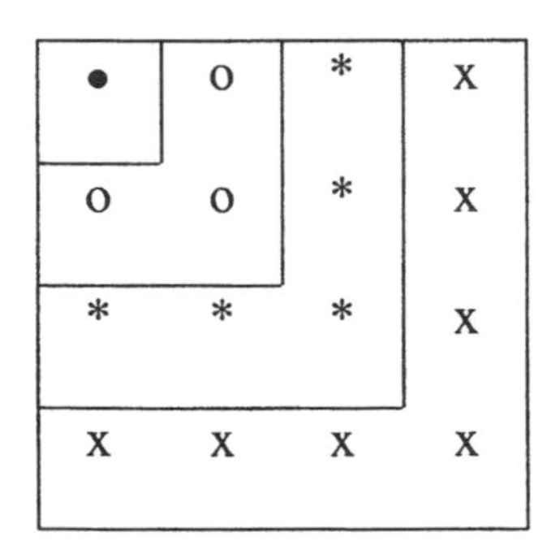

1

$1 + 3 = 2^2$

$1 + 3 + 5 = 3^2$

$1 + 3 + 5 + 7 = 4^2$

This action proof belongs to the *diknumi* tradition of ancient Greece. A *diknumi* proof (Fauvel, 1987) is one that is presented—typically by means of a diagram of some sort—in such a way that no explanation is necessary, for one can "see" the result and the argument. For example, a suitable arrangement of pebbles in pairs "demonstrates" that the sum of two odd numbers is even. The example displayed—the arrangement of a particular set of pebbles—is a generic example, in that it *points* to a more general truth. In the *diknumi* proof, a train of thought is shared yet unspoken. An extensive and varied collection of such "proofs without words" is assembled in Nelson (1993).

Divisibility by 9

In a report of their study of college students' proof schemes, Harel and Sowder (1998) describe the approach taken by several students to the proof of the digit-sum test of divisibility by 9, by

> taking a specific whole number, say 867, and saying something to the effect: This number can be represented as $8 \times 100 + 6 \times 10 + 7$, which is $(8 \times 99 + 6 \times 9) + (8 + 6 + 7)$. Since the first addend, $8 \times 99 + 6 \times 9$, definitely is divisible by 9, the second addend, $8 + 6 + 7$, which is the sum of the number's digits, must be divisible by 9. Some of the students indicated, in addition, that this process can be applied to any number. (p. 271)

Harel and Sowder comment that "This scheme reflects students' inability to express their justification in general terms . . ." (1998, p. 271). I would, however, beg to differ with them on this issue. It all depends on what the students imagined their proof was *for*. If the supposed purpose was *explanation* (see Hersh, above), then they may have thought that one illuminating example, carefully analyzed, was the most effective way to make their case. I return to this issue toward the end of this chapter.

Gauss and the Sum 1 + 2 + 3 + . . . 100

The story (probably apocryphal, but see Polya, 1962, pp. 60–62, for one version) is told about the child C. F. Gauss, who astounded his village schoolmaster by his rapid calculation of the sum of the integers from 1 to 100. While the other pupils performed laborious column addition, Gauss added 1 to 100, 2 to 99, 3 to 98, and so on, and finally computed fifty 101s with ease. The power of the story is that it offers the listener a means to add, say, the integers from 1 to 200. Gauss's method demonstrates, by generic example, that the sum of the first $2k$ positive integers is $k(2k + 1)$. Nobody who could follow Gauss's method in the case $k = 50$ could possibly doubt the general case. It is important to emphasize that it is not simply the *fact* that the proposition $1 + 2 + 3 +$

$\ldots + 2k = k(2k + 1)$ has been verified as true in the case $k = 50$. It is the *manner* in which it is verified, the form of presentation of the confirmation.

Drawing on E. T. Bell's classic volume (1937), Paul Hoffman (1998) recounts the story in his recent best-seller *The Man Who Loved Only Numbers*. His comment on it (quoting mathematician Ronald Graham) is a telling testimony to the genericity of Gauss's method: "What makes Gauss's method so special. . . . Is that it doesn't just work for this specific problem but can be generalised to find the sum of the first 50 integers or the first 1,000 integers . . . or whatever you want" (p. 208).

I should add that, in introducing the notion "generic example" to audiences of all kinds—undergraduate and graduate students, mathematics education conference-goers, "general audiences"—I routinely choose Gauss's method as a paradigm of the genre: generic among generic examples.

Euler's ϕ-function is Multiplicative

For any natural number n, $\phi(n)$ is defined to be the number of natural numbers less than and co-prime[7] to n. Thus, denoting the number of elements in the set A by #(A), $\phi(10) = \#\{1, 3, 7, 9\} = 4$. Examples readily lead to the conjecture that $\phi(mn) = \phi(m)\phi(n)$ when m, n are co-prime. Walther (1986) gives a generic argument for the case $m = 3$, $n = 8$.

Assign the integers 0 to 23 to the cells of a matrix with 3 rows and 8 columns, corresponding to the 3 residue classes modulo 3 and the 8 residue classes modulo 8, respectively.[8] The fact that there will be one and only one integer in each cell is a consequence of the Chinese Remainder Theorem.[9] It "works" because 3 and 8 are co-prime.

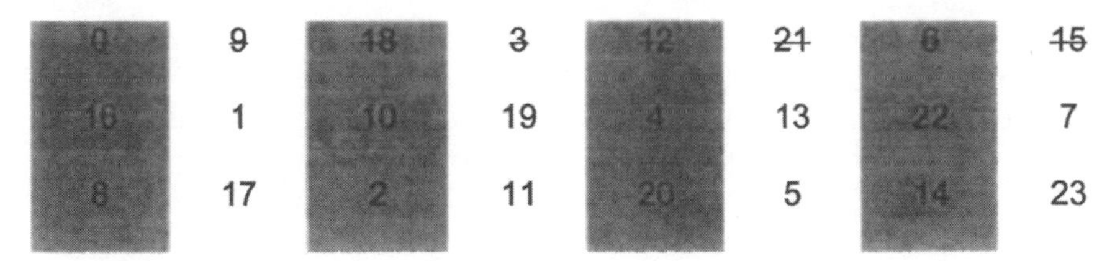

0	~~9~~	18	~~3~~	12	~~21~~	6	~~15~~
16	1	10	19	4	13	22	7
8	17	2	11	20	5	14	23

Now delete each element, which is not co-prime with 8. (These elements have been shaded, e.g., 16.) Observe that these form whole columns, for d divides t and 8 if and only if d divides $t \pm 8$ and 8. Note that $\phi(8) = 4$ columns remain undeleted (i.e., unshaded in the diagram), corresponding to those integers from 1 to 7, which are co-prime with 8 (no two of which can be in the same column). Next, delete each element, which is not co-prime with 3 (shown as strike-through, e.g., ~~9~~). As before, if an element is deleted, every element in the row containing it is also deleted, and $\phi(3) = 2$ rows remain undeleted. It is apparent, therefore, that the total number of elements in the undeleted array with $\phi(8)$

columns and $\phi(3)$ rows is the product of $\phi(8)$ and $\phi(3)$. Finally, we note that the undeleted elements are precisely those co-prime with 24. This is no coincidence, for $t \in N$ is co-prime with 24 if and only if it is co-prime with both 3 and 8. (This claim is logically equivalent by negation to the statement: $t \in N$ shares a nonunit factor with 24 if and only if it is not co-prime with 3 or 8).[10]

At no point in this argument do we draw on any particular properties of 3 and 8, other than the fact that they are co-prime. The argument and the conclusion would transfer to any co-prime pair (m, n).

Constructing Castles

I have a certain fondness for a recently rediscovered article by Larry Copes (1980). While on leave of absence from his college duties, Copes makes time to revisit some number theory that he had known—even "passed a test on"—as a student. He chances on a theorem due to Euler:

$$\text{For every } n \in N, \sum_{d|n} \phi(d) = n \text{ (taking } \phi(1) \text{ to be 1)}$$

Most of the article concerns Copes's trying to make sense of this theorem. He vaguely recalls "having used it in graduate school to prove something else" (p. 44).[11] He finds and reproduces the textbook proof of Euler's Theorem, then proceeds to attempt to fit the proof onto his "weird" and complex "castle" of mathematical knowledge by reworking it with reference to a special case: $n = 12$. "Well, I thought, let's take one sentence at a time, using an example. What is a good example? Nothing too special—how about 12?" (p. 44).

What was Copes's sense that 12 is "nothing too special"? We can speculate that 12 is not prime. A prime value p of n would not "speak the generality" of the proof, since $\Sigma\phi(d)$ would then trivially reduce to $\phi(1) + \phi(p) = 1 + (p-1) = p$. By contrast, 12 has more than two divisors, but not too many. At one point, Copes is trying to see *why* it turns out that, for all divisors d of 12, $\#\{m \in N: m < 12 \text{ and } \gcd(m, 12) = d\} = \phi(12/d)$. He comments:

> What was I really asking? I was not asking to be convinced, because a couple of examples had persuaded me. Nor was I asking for a logical proof. . . . No, I was looking for some deeper relationships among those numbers. I was looking for more understanding, more meaning. . . . (p. 44)

Copes's remark about "a couple of examples" nicely foreshadows Hersh's whimsy (above) about students being all too easily convinced. But he is not content with knowing (or believing) *that*, he wants to know *why*, and a formal proof from a well-respected textbook does not meet his need. The spirit of Copes's exploration is very much in keeping with the nature and epistemological purpose of a generic example.

GENERIC PRESENTATION OF PROOFS IN NUMBER THEORY

The examples of number-theoretic generic proofs given earlier are useful up to a point, but how might we "compose," write, devise (or whatever) such proofs for ourselves and/or for our students? The following brief discussion of certain possibilities and limitations of generic approaches to some well known theorems is a contribution to the search for an answer to this question.

Wilson's Theorem

Kate's proof that square numbers have an odd number of divisors is reminiscent of the standard proof of Wilson's Theorem, that $(p - 1)! \equiv (p - 1)$ modulo p, for all primes p. What might a generic proof of that theorem look like? As a preliminary, we would need to know that ± 1 are the only self-inverse elements under multiplication modulo p. I concede, if that is the right word, that while this result emerges readily (as a conjecture, of course) from examples, especially when the contrast is made with nonprime moduli, its proof[12] does not seem to lend itself to generic presentation. Now consider the prime number 13 (17 or 19 would do equally well) and list the reduced set of residues[13] modulo 13:

1 2 3 4 5 6 7 8 9 10 11 12

Pair each of the numbers from 2 to 11 with its (distinct) multiplicative inverse mod 13, to give (2, 7), (3, 9), (4, 10), (5, 8), (6, 11), with 1 and 12 being self-inverse. (I usually link the elements in the inverse-pairs with lines on a chalkboard.) Clearly, the product of these integers from 2 to 11 must be congruent to 1^5, that is, 1, modulo 13. Therefore $12! \equiv 1 \times 1 \times 12$ (= 12) mod 13. The argument is generic, since 13 was in no way an untypical choice: the pairing would work equally well with any prime. An account of how this worked with one undergraduate class is given later in this chapter.

The Set of Primes is Infinite

Euclid's classic proof that the set of prime numbers is infinite can be presented either as a proof by contradiction, or as a constructive argument.[14] In the first case, we suppose that there is only a finite number, and list them as $p_1, p_2, \ldots, p_n$. The integer $N = p_1 p_2 \ldots p_n + 1$ is then clearly not a multiple of any of the primes listed, yet (being an integer) it must have a prime divisor, so we have a contradiction. The constructive argument proceeds as follows. Let $p_1 = 2$ (this is arbitrary, it could be any prime), and for every natural number n, define p_{n+1} to be a prime (the least, say), which divides $N = p_1 p_2 \ldots p_n + 1$. The existence of p_{n+1}. distinct from $p_1, p_2, \ldots, p_n$ needs to be justified as before. This recursive definition ensures a countably infinite set of primes.

The proof by contradiction cannot be presented generically, because—in the opening gambit, "suppose not"—one cannot give a specific example of something that is not the case. On the other hand, the constructive proof lends itself very well to generic exposition. One could invite the class to pick any prime as a starting point, $p_1 = 7$, for example. Then $p_2 = 2$, being the least prime divisor of $7 + 1$, $p_3 = 3$ (the least prime divisor of $7 \cdot 2+1$) and the sequence (p_n) proceeds 7, 2, 3, 43, 13, ... With $p_1=5$, we find $(p_n) = 5, 2, 11, 3, 331, \ldots$. The generic argument not only makes apparent the inevitability of a new prime at every stage, but offers the satisfaction of actually identifying it (especially if suitable software is at hand to factorize large integers). There seems to me to be a human preference for "closure" (Collis, 1972) when mathematical objects are being manipulated: knowing not merely that something exists, but knowing what it is. The generic proof (above) that no finite set can ever exhaust the primes seems to meet this human preference.

Gauss's Lemma

Two preliminary definitions may be helpful here.

- Let p be a prime number. An integer a is a quadratic residue of p if $x^2 \equiv a \bmod p$ for some integer x in the range $1 \leq x \leq p - 1$. Thus 3 is a quadratic residue mod 11 because $6^2 \equiv 3 \bmod 11$.
- The least absolute (complete) set of residues modulo p is the set $\{0, \pm1, \pm2, \pm3, \ldots, \pm\frac{1}{2}(p-1)\}$. (The term "least *absolute*" serves to distinguish this set from the least *positive* set of residues [note 13]. Every integer is congruent modulo p to just one of least absolute residues. The integers $-1, -2, -3, \ldots, -\frac{1}{2}(p-1)$ are referred to as the negative least absolute residues mod p.)

One of the fundamental problems in the study of quadratic residues is determining whether or not a given integer a is a quadratic residue of a given prime p, without recourse to direct calculation of the "squares" modulo p. One somewhat "bare knuckle" answer is provided by Euler's Criterion, which asserts that a is a quadratic residue of p if and only if $a^{\frac{p-1}{2}} \equiv 1 \bmod p$. This is of limited practical use unless a and p are relatively small. A more ingenious solution is Gauss's Lemma, which states that[15] a is a quadratic residue of p if and only if, in the set $a, 2a, \ldots \frac{1}{2}(p - 1)\, a$, the number of negative least absolute values modulo p is even.

For a generic proof (and, hopefully, some unpacking of what the Lemma is saying), consider the case when $p = 19$ and $a = 3$. We shall compute the 9 products 3•1, 3•2, . . . ,3•9, and assign each in turn to the corresponding least absolute residue to which it is congruent, modulo 19. These least absolute residues are ±1, ±2, ±3, ±4, ±5, ±6, ±7, ±8, ±9. They are displayed in the first and third rows of the table below, with the corresponding values of 3•*k* shown in the middle row.

1	2	3	4	5	6	7	8	9
3•6	3•7	3•1	3•5	3•8	3•2	3•4	3•9	3•3
-1	-2	-3	-4	-5	-6	-7	-8	-9

In each case, the value of 3•*k* is shaded together with its least absolute value mod 19, above or below it. (In class exposition, this would be performed actively, with the involvement and assent of the class.) For example, in the first column, 3•6 is one fewer than 19, so 3•6 is shaded with −1 below it.

No ambiguity arises in the shading. For example, 3•4 is associated with −7 (because 12 ≡ −7 mod 19) and not with 7. This is to be expected—the set 0, ±1,±2, . . . ±9 is a complete set of residues, with no redundant elements. We might note, with some satisfaction, that the 9 multiples of 3 are assigned one–one to the middle row of cells, with no double entries or omissions. On reflection, this is to be expected, for if $1 \leq k_1 < k_2 \leq 9$ with $3 \cdot k_1 \equiv r \bmod 19$ and $3 \cdot k_2 \equiv -r \bmod 19$, then $3 \cdot (k_1 + k_2) \equiv 0 \bmod 19$, hence $k_1 + k_2 = 19$. But $k_1 + k_2$ is at most 17.

The elements of each shaded pair are congruent to each other mod 19, so their products must also be congruent mod 19. Thus:

$(3 \cdot 1) \times (3 \cdot 2) \times \ldots (3 \cdot 9) \equiv (-1) \times (2) \times (3) \times (-4) \ldots (9) \bmod 19$, and so
$3^9 \times 9! \equiv (-1)^3 \times 9! \bmod 19.$

Now 9! is co-prime with 19, so we may conclude that

$$3^9 \equiv -1 \bmod 19.$$

It now follows from Euler's Criterion that 3 is not a quadratic residue of 19.

GENERAL PRINCIPLES

While it is premature to suggest a definitive list of principles underpinning the construction and presentation of such generic proofs in number theory, it is possible to offer a small number of guiding principles, building on the discussion in Walther (1986, pp. 265–267).

1. The particular case should be neither trivial nor too complicated (Semadeni, 1984). If the proposition to be proved concerns a property of prime numbers, the range $13 \leq p \leq 19$ seems very suitable, but the subsequent argument must determine the choice. For example, if we are to discuss factors of $p - 1$, we

might avoid 17 because every factor of 16 is a power of 2. If the proposition concerns natural numbers in general, don't choose a prime for the generic example.

2. Suppose (as is typically the case) we are to prove a statement of the kind "for every natural number n, $A(n)$." In a generic proof, we begin with a particular value n_0 of n, and demonstrate $A(n_0)$ in such a way that analogy with other instances is readily achieved. For this transfer to be possible, the value n_0 needs to be transparent throughout the proof; we need to be able to "track"[16] it through the stages of the argument. This is necessary not only for the genericity of the argument, but to enable the transition (if desired) to formal, general proof.
3. Thus in providing and presenting insight as to why $A(n)$ holds for a particular n_0, emphasize aspects of the proof that are invariant regarding a transfer to other (arbitrary) values of n. Such invariants are primarily constituted in *relations between* mathematical objects. That is to say, they are rooted not in the mathematical objects (such as n_0) themselves, but in operations on such objects (Walther, 1986).
4. The reasoning should be constructive. That is to say, where it is argued that certain mathematical objects (usually natural numbers) exist, they should, when practicable, be identified and verified to have the properties claimed for them. At the same time, it is beneficial to recognize and anticipate situations when this may be potentially misleading. For example, in Euclid's proof that the set of primes is infinite, it is commonplace for students to suppose (notwithstanding careful exposition about prime *factors*) that $N = p_1p_2 \ldots p_n + 1$ is itself always prime. The generic counterpart initially reinforces this misconception, with $N = 3, 7, 31, 211, 2311$ for $p_n = 2, 3, 5, 7, 11$, respectively. Only when $p_n = 13$ do we encounter the composite $N = 30031 = 59 \times 509$.
5. For novice students, the transition (if desired) to formal, generalized argument is not a formality. Initially, some scaffolding is needed to ensure that students perceive the invariants of the argument, and can summon up the *notation* needed to communicate them. In time, students are likely to become more independent in this respect. See, for example, the case of Jonathan (Rowland, 1998).

Caveat

However convincing, or otherwise, we might find the generic proofs above (and others) for ourselves, we ought to bear in mind the caveats expressed by some writers. Mason and Pimm (1984), for example, write: "A generic example is [. . .] presented in such a way as to bring out its intended role as the carrier of the general. This is done by stressing and ignoring various key features. [. . .] Unfortunately it is impossible to tell whether someone is stressing and ignoring in the same way as you are" (p. 287).

While this must be true, it should not be cause for despair—certainly not cause for abandoning generic arguments, or for believing that a conventional, formal proof meets the student's desire (or need) to know *why*. Semadeni (1984) comments: "An action proof [. . .] involves a psychological question: how can one know whether the child is convinced of the validity of the proof . . . ? Without dismissing this criticism, we note that it applies to any proof in a textbook: if the author finds his proof correct and complete, this does not automatically imply that students understand it" (p. 34). We come now to some evidence, which suggests that many (or most) students—but not all—may indeed be stressing or ignoring in the way we intend

WORKING WITH UNDERGRADUATES

My purpose in this section is to indicate ways that I have worked generically with undergraduate students in courses on number theory, applying the speculation of the previous sections to the real world of the classroom. In addition, I shall give some empirical data, albeit small scale. The intent of these accounts is to focus on what students "see" in generic examples, on the issue of transfer to other examples, and on the transition to formal proof. The narrative is chronological, reflecting my own awarenesses, and the priority I attached to particular concerns over time.

The students in question are following a four-year mathematics/education program that leads to a Bachelor of Education (B.Ed.) degree and qualifies them as generalist elementary school teachers with specialist expertise in mathematics. This program is one of a number of undergraduate routes to Qualified Teacher Status[17] (QTS) in England. It is true to say that the Cambridge B.Ed. undergraduate program for QTS places unusually high emphasis on study of the specialist discipline, mathematics in this case, and the students are unusually well-qualified for QTS undergraduates.

Academic mathematics courses take place in the first two years of the B.Ed. program, the final two years focusing on pedagogy, education disciplinary studies, and placements in schools. These mathematics units are taught to classes of twenty–twenty-five, supplemented by weekly "supervision" classes in groups of three, at which student difficulties with lectures and problem sheets are addressed. In Cambridge, the mathematics units for the B.Ed. are in algebra and geometry (the application of groups and vector spaces to transformation geometry in two and three dimensions), number theory, numerical and discrete mathematics, and in probability. This advanced mathematics curriculum is designed and taught by mathematics educators, being chosen for its relevance to elementary mathematics, in order to unify academic study of mathematics and the students' professional goals.

Primitive Roots

For some years I taught the thirty-six-hour second year course in number theory, covering the usual topics found in texts such as Baker (1984). Modular arithmetic arises early in the course. Again, it may be helpful here to introduce some ideas and terms referred to later in this section.

- For any natural number n, M_n will denote the group whose members are the reduced set of residues modulo n (refer to note 13). The group operation is multiplication modulo n.
- A multiplicative group G is *cyclic* if there exists an element a of G such that every element can be expressed as an integer power a. In this case, a is said to be a generator of G.
- In general, M_n is not cyclic, for example, in $M_8 = \{1, 3, 5, 7\}$, the powers of each element are either itself or 1, so no single element generates all four elements of M_8.
- On the other hand, $M_{10} = \{1, 3, 7, 9\}$ is cyclic, being generated by 3 (and also by 7).
- If M_n is cyclic, any generator is called a *primitive root* of n.

One major result in this area is that every prime number p has a primitive root; that is to say, the group $\{1, 2, 3, \ldots, p—1\}$ under multiplication modulo p is cyclic. Unfortunately, confirming instances of this theorem give no clue as to how it might be proved. The standard general proof (see, for example, Baker, 1984, p. 23) is surprisingly indirect and overburdened with notational complexity. This complexity is, in part, due to the fact that one has to deal with two universally quantified variables (any prime p and any divisor d of $p - 1$), resulting in a double layer of generality.

For the first few times that I gave this course, I "taught" this general proof and applied all my powers of exposition and explanation to make it accessible to students. I noted their difficulties and returned to my notes to fine-tune and polish my presentation for the following year. I filled in and amplified the steps of the argument, explaining many minute details. While my effort was well-intended, it was a case of self-delusion: deep down, I knew that I was adding to my own appreciation of the proof, but that the improvements were at best marginal from the students' point of view. Worse, the atomizing of the subtleties of the general argument into myriad tiny steps left students with no clear overview of the grandeur of the design.

One year, confronted with rows of puzzled faces at the completion of my exposition, desperation inspired me to add, "Well look, suppose p is 19, so that the order of M_p is 18. Now, what are the possible orders of the elements of M_p?" I continued, and as I did so it dawned on me that my choice of this particular example as a vehicle for the proof was enabling me to give a faithful account of the general argument. I soon began to theorize about that process, and learned to call it proof by generic example. The next year, and the next, I began my exposition of the primitive root theorem with the generic example, and challenged the students to "go home and write out the general proof." Thereafter, I began to wonder why I was asking them to do that, since they would only be able to do so if their "understanding" of the generality of the argument (borne by the generic example $p = 19$) was as complete as I could wish it to be.

Eventually, I carried out an enquiry with one class of students following this number theory course program. The generic argument is summarized here. It omits many details that are considered in my classroom exposition but that are not essential to its consideration as a generic proof.

THE PRIME NUMBER $p = 19$ HAS A PRIMITIVE ROOT

1. The group M_{19} has 18 elements, so the order of each of those elements must divide 18.
   Possibilities are 1, 2, 3, 6, 9, and 18.[18]
   Suppose there are N_1 elements of order 1, N_2 of order 2, . . . , N_{18} of order 18, so $N_1 + N_2 + N_3 + N_6 + N_9 + N_{18} = 18$. To prove that 19 has a primitive root,
   we need to demonstrate that $N_{18} \neq 0$.
2. Focus for the moment on the elements of order 6 (there might be none). Argue (as in the lecture) that if $N_6 \neq 0$, then $N_6 = \phi(6)$. A similar line of reasoning would establish that $N_1 = 0$ or $N_1 = \phi(1)$, $N_2 = 0$ or $N_2 = \phi(2)$, . . . , $N_{18} = 0$ or $N_{18} = \phi(18)$. It follows that $[\phi(i) - N_i] \geq 0$ for each of the 6 possible values of i.
3. We know that $\sum_{d \backslash n} \phi(d) = n$ for all $n \varepsilon N$,
   so $\phi(1) + \phi(2) + \phi(3) + \phi(6) + \phi(9) + \phi(18) = 18$.
   Since $N_1 + N_2 + N_3 + N_6 + N_9 + N_{18} = 18$, it follows that
   $[\phi(1) - N_1] + [\phi(2) - N_2] + [\phi(3) - N_3] + [\phi(6) - N_6] + [\phi(9) - N_9] + [\phi(18) - N_{18}] = 0$.
   Since each bracket is non-negative, they must all be zero.
4. In other words, $\phi(1) = N_1$, $\phi(2) = N_2$, . . . and in particular, $\boldsymbol{\phi(18) = N_{18}}$.
   Now $\phi(18) = 6$, so it follows that **19 has a primitive root**—6 of them, in fact.

The students were asked to make a written response to three questions:

1. Are you happy with the above explanation for the case $p = 19$, that is, is it convincing?
2. Does the above explanation help to convince you that 29 has a primitive root?
3. Does the particular explanation (for the case $p = 19$) convince you that *every* prime has a primitive root?

These students were invited to elaborate on their responses if they felt able to do so. The 19 questionnaire returns indicate that the argument concerning the case $p = 19$ was generic for 12 of the students, whose responses included the following:

> [concerning $p = 29$] The whole process could be repeated using 29 so that 1, 2, 4, 7, 14, 28 are possible orders, for example 7, same argument as for M_{19}. Come to same conclusions, you just have different numbers involved.

> I went through the proof with $p = 29$ and felt that it was applicable.

> You can adapt the proof so that it would apply to the possible orders of M_{29}, and then follow through the same argument.

> M_{29} has 28 elements with divisors 1, 2, 4, 7, 14, 28 [. . .] $N_1 + N_2 + \ldots N_{28} = 28$, similarly $\phi(28) = 28$ because of the theorem [. . .] I can see that the argument can be transferred to $p = 29$ and would show that $N_{28} = \phi(28) = 12$. Therefore, 29 has 12 primitive roots—quite convincingly!

> [concerning any prime] It is easy to follow the logical progression of the proof for $p = 19$ with any other prime in mind, and I can see no area of the proof that gives me any doubt that it wouldn't work for any prime.

Some respondents stated that they appreciated the "concreteness" of the generic argument: "If it was general with no numbers, I think I would get confused." Two students, moreover, volunteered that the generic argument with $p = 19$ indicated (for them) how they might formulate a general proof:

> Because you could just extend the argument for any prime and substitute in the values, and perhaps produce a general form for the proof.

> By changing 19 to p, I could generalize the argument.

For four of the students, the generic intention of the example was not effective. They were united by the sense that the proof for $p = 19$ was precisely that, and no more:

> Although the explanation for $p = 19$ is clear and true, it doesn't necessarily follow that $p = 29$ has a primitive root. So I'd prefer to work through it [$p = 29$] before I was convinced.

In effect, these students are being cautious about what they perceive as a case of empirical generalization (Bills, 1996). They fail to see that the argument has been presented with the aim of *structural* generalization; rather, the argument is not effective in suggesting structural generalization to them. It could be, of course, that these four second-year undergraduate students are conditioned into being satisfied with nothing less than a "general" algebraic argument: indeed, they all indicated their need for a "general proof" before they could be convinced. One student spoke for all four when he wrote: "I would like a general proof to reassure me for all primes." Although he added somewhat wistfully: "However, a general proof on its own would probably confuse me."

The ambivalence of the three remaining responses could not be resolved, since all the returns were anonymous. On the whole, these respondents found the case for primes other than 19 to be plausible but not entirely convincing. The distinction between public accountability and personal conviction is brought out in the following comment:

> It needs to be proved generally [. . .] but using a number 19 makes it easier to follow. [. . .] A general proof needs to be followed through step by step. (But I am fairly convinced already—I can't see why it wouldn't work.)

As Walther (1986) indicates, the psychological effectiveness of the generic proof hinges on the identification and transfer of structural invariants, as opposed to contingent variables, in the argument. The sophisticated mathematician is able to isolate such invariants; indeed, some of the quotations from the students' writing above bring this out explicitly, such as: "You can *adapt* the proof" or "The argument can be *transferred*." One student uses the word *similarly* to refer to this transfer from one particular case to another.

The Euler Function

The context now shifts to a class of first-year Cambridge B.Ed. undergraduate students attending a short "taster" module in number theory. The background of these students is indicated earlier in this chapter. Their responses to one question in the mid-course assignment were informative and salutary. This question asked:

(i) Draw a Venn (or Carroll) diagram with universal set $E = \{1, 2, 3, \ldots, 34, 35\}$ and subsets $F = \{\text{multiples of } 5\}$, $S = \{\text{multiples of } 7\}$. State the number of elements in each of these two subsets, and the number in their union. How would you characterize the elements in the complement of $F \cup S$?

(ii) By generalizing (i), prove from first principles that if p,q are distinct primes, then $\phi(pq) = (p - 1)(q - 1)$.

My intention, in devising the question, was to elicit solutions of the following kind (the versions given here are intentionally succinct).

(i) $\#(F) = 7$, $\#(S) = 5$, $\#(F \cup S) = 7 + 5 - 1 = 11$. $(F \cup S)'$ is M_{35}—the set of natural numbers less than and co-prime with 35.

(ii) Let $E = \{1, 2, 3, \ldots, pq\}$, with $P = \{\text{multiples of } p\}$, $Q = \{\text{multiples of } q\}$. Then $\#(P) = q$, $\#(Q) = p$, $\#(P \cup Q) = p + q - 1$. $(P \cup Q)'$ is

M_{pq} and so $\phi(pq) = \#(P \cup Q)' = \#(E) - \#(P \cup Q) = pq - (p + q - 1) = pq - p - q + 1 = (p - 1)(q - 1)$.

Each of the twenty students gave good or adequate solutions to the first part, but only ten seemed to see it as generic in relation to the second. Three seemed not to comprehend what was expected in terms of "first principles," and drew on "big" theorems such as $\phi(pq) = \phi(p)\phi(q)$. This is a separate issue. What was more interesting (a teacher's euphemism for "worrisome"?) was the handful who argued as follows.

> In the first part, $(F \cup S)'$ is the set of natural numbers less than and co-prime with 35, and in that case $\phi(5 \times 7) = \#(F \cup S)'$ turns out to be 24. Now 24 is equal to $(5 - 1)(7 - 1)$ so, in general, $\phi(pq) = (p - 1)(q - 1)$.

One student even said that the general result followed because the case for 35 was a generic example. Two thoughts come to mind. First, that the transition from $\phi(5 \times 7) = (5 - 1)(7 - 1)$ to $\phi(pq) = (p - 1)(q - 1)$ seems to be a case of empirical rather than structural generalization, in that the general statement mimics the *form* of the particular, in the absence of explanatory power. In the case of this exercise, of course, the intended generalization is provided for the student, who then looks for an instance of it in the (particular) first part. My second thought is that the example in the first part of the question is incomplete in the generic sense. The second part seems to *require* some symbol manipulation to show that $pq - p - q + 1$ and $(p - 1)(q - 1)$ denote the same function. Nevertheless, for half of the students, awareness of the reasoning process in the first part with 5 and 7 enabled them to construct a formal proof of the general case for arbitrary, distinct primes p and q. I would regard that outcome as encouraging, given that the issue of transition from the generic to the general had not been explicitly addressed at this stage of their course. This issue is revisited in the next subsection.

Wilson's Theorem Revisited

More recently, I "taught" Wilson's Theorem to the same class of first-year undergraduate students. It took an hour. Of course, I could have stated the theorem and proved it formally in five minutes. Instead, I asked them to evaluate 4! mod 5, 6! mod 7, 10! mod 11, and to write down a conjecture. The modal version of the conjecture was $n! \equiv n \bmod (n + 1)$. The "for all n" seemed to be implicit. I asked them to evaluate 5! mod 6. They did, and they were visibly surprised by the refutation. I asked whether they could modify the conjecture. At first they honed in on the even/odd distinction between moduli, but $n = 8$ led to further refutation and eventual restriction to prime values of $n + 1$. Then $n = 12$ provided a further confirming instance. I proceeded to an interactive presentation of a generic proof,[19] inviting Sonia to pick a prime between 11 and

19. She chose 19. I got them to list 1 to 18 and work on inverse pairs in groups, during which Simon spontaneously explained to his colleagues why 18! *had* to be 18 mod 19. I asked him to repeat his reasoning to the class, and wrote his explanation on the whiteboard. They dutifully copied it. Later, I inquired what would have happened if we had looked at 28! mod 29, and Abby explained why it would have to be 28. "Does everyone agree?" I asked. They agreed. One shouldn't read too much into such consent, however pleasing; sometimes, they just want to get a good place in the queue for lunch. Nevertheless, Abby, at least, had convinced me that she had appropriated the generic proof.

The next day, at a tutorial meeting, I asked five members of the class to write out the proof—that, for primes p, $p! \equiv p - 1 \bmod p$—in conventional generality. Their responses were unaided and individual. These students will have had little experience of composing formal proofs. Nevertheless, they all indicated in their writing that the genericity of the case $p = 19$ had been apparent to them. Moreover, their argumentation and use of notation would have satisfied any examiner. Hannah's response, which was typical, was as follows.

> $(p - 1)(p - 2)(p - 3)(p - 4) \dots 2 \times 1$
>
> Every element of M_p[20] has an inverse, because M_p is a group.
>
> We know (from work on primitive roots) that only $p - 1$ has order 2. Therefore, $p - 1$ is self-inverse. All other members of M_p apart from 1 must have a distinct inverse.
>
> Each inverse pair when multiplied gives 1 mod p.
>
> This gives $(p - 1)(1^{1/2(p-3)})\ 1 \equiv (p - 1)! \bmod p$
>
> Therefore, $(p - 1)! \equiv (p - 1) \bmod p$

Only Zoë gave evidence of some insecurity in this intangible world that lies beyond examples. Her proof was much the same as Hannah's, but began with identification of the inverse pairs in the case $p = 11$ (transfer to other examples) and concluded the comment: "I tried to find a formula for the inverses, for example $p - 2$ has inverse $p - 6$ (but only for $p = 11$). I have been unable to do this."

For Zoë, mere knowledge of the *existence* of distinct inverses in the range 2 to $p - 2$ is not enough. What is not clear is whether it leaves *her* cognitively insecure, or whether she believes that I (in my role as assessor) will expect more. Abby's proof was elegantly and lucidly expressed, but stated that there are $\frac{1}{2}(p - 1)$ inverse-pairs rather than $\frac{1}{2}(p - 3)$—an error of manipulation, but not one of conception.

DISCUSSION

I have identified three areas for summary discussion, to offer some pointers for "expert" pedagogical deployment of generic examples as a means whereby students might meaningfully engage with number theoretic arguments and consequently experience some personal stake in the generality of proof. First, I revisit the examples and the "guiding principles" proposed earlier. Second, I consider what can be learned from reflection on my work with undergraduate students. Finally, I make some pedagogical suggestions, with a view to enabling students to "see" the general arguments embedded in particular instances. Of course, these three areas are not distinct: discussion of some issues will drift between the three subsections that follow.

Examples and Principles

The examples given here of generic proof, from the literature and those constructed for my own teaching, consist of attempts to convey to the student (or the reader) some insight that extends way beyond the case in point. In most cases, they seem to derive from the authors' attempts to make sense—for themselves, or for their audience—of a conventional, formal argument.

That argument is embedded and faithfully rehearsed in a special case, with regard for the structure of the argument, endeavoring to facilitate the identification and transfer of paradigm-yet-arbitrary values and structural invariants within it (Principles 2 and 3). For example, the generic exegesis of the test for divisibility by 9 explicitly reveals each power of 10 to be one more than a multiple of 9, and isolates both the sum of the digits and a multiple of 9. In this way the digits are tracked through the proof, and seen to be present when separated off from the 9s. Similarly, the summation of $1 + 2 + 3 + \ldots + 2k$ through the case $k = 50$ emphasizes the first–last pairings, the invariance of the sum in each case, and the inevitability of k such pairs. The exposition of Gauss's Lemma in the case $p = 19$, $a = 3$ displays the multiples of 3 as products, that is, $3 \cdot 1, 3 \cdot 2 \ldots 3 \cdot 9$, and not as realized values $3, 6, \ldots 27$. The proof of Wilson's Theorem stresses the guarantee of the multiplicative inverse pairings within the range 2 to $p - 2$. This nicely exemplifies the ideal that invariants (here, the inverse pairings) are constituted not in mathematical objects (the integers 1 to $p - 1$) but in relations between them (the pairings).

The choice of a particular case is a delicate matter, but Principle 1 seems to offer a sensible starting point for making such choices. In the Jailer problem, Kate's choice of 36 is interesting—small enough to be accessible with mental arithmetic but with sufficient factors to be nontrivial. By contrast, 25 or 49, being squares of primes, would yield a more impoverished and less paradigmatic set of factors; 36 is, in fact, the smallest square with more than three factors. Recently, I set the following task for a seminar audience,[21] in advance of viewing the *Teachers Count* video clip:

> The well-known "Jailer" investigation may lead to the observation (conjecture) that square numbers have an odd number of divisors.
>
> 1. Discuss this conjecture with a colleague.

2. Devise a generic example for the purpose of explanation to a pupil about 12 years of age.

One participant, Alison, specializes in elementary school mathematics. Like Kate, she chose 36 for the purposes of her generic example, and her generic reasoning about the factor-pairs uncannily anticipated what we subsequently saw in the video.

Regarding Principle 4, the constructive presentation of generic proofs, for the moment I have little to add to what Uri Leron (1985) has already written in connection with formal proofs. Leron points to a fundamental psychological difficulty associated with proof by contradiction. Most nontrivial proofs, says Leron, pivot around the construction of a new mathematical object (number, function, line, etc.), offering the Platonic sense that we are acting on "reality." In indirect proofs, by contrast

> we begin the proof with a declaration that we are about to enter a false, impossible world, and all our efforts are directed towards destroying this world, proving it is indeed false and impossible. . . . Actually, there is a way to alleviate the frustration . . . in many indirect proofs, the main construction is independent of the negative assumption. You can therefore *separate out the construction from the negative assumption* . . . . (pp. 323–324)

This human preference for construction is related to, though not identical to, the preference for "closure" to which I referred earlier. Leron (1985) speaks of the "cognitive strain" set up in the learner who is denied the possibility of grounding argument in, or making contact with, mathematical objects that can readily be recognized as "real" and tangible.

The transition to formal argument (Principle 5) does seem to be facilitated by first considering explicitly the transfer to other examples (Principle 3). It is easier, as a first step, to clarify first for oneself, then to communicate to others, what is perceived in one example by reference to another. This is tied up in the intimate relationship between prediction and generalization, which I have discussed elsewhere (Rowland, 1999). These points are developed further in the next subsection.

Undergraduate Students: Trends and Obstacles

The idea that elementary school students might be enabled to gain insight into the "why" of certain mathematical generalizations through examination of well-chosen (and well-structured) examples seems relatively uncontroversial. This is reflected in the title of Semadeni's 1984 paper: "Action Proofs in Primary Mathematics Teaching and in Teacher Training," in which he writes:

> As a rule, proofs are not used in primary teaching. The reason for this is obvious: the concrete-operational child is not capable of hypothetical reasoning, of deduction expressed in words and symbols. . . . The purpose of this paper is to propose a general scheme for devising primary-school proofs, called action proofs. (p. 32)

But what attitude should we adopt toward the use of generic proofs in advanced mathematics, and in number theory in particular? I believe that the accounts given here of my work with undergraduates offer grounds for considerable optimism regarding the possibility of students "seeing" the generality we intend them to see in arguments based on particular cases. At the same time, it warns us against naive complacency: we cannot be sure what they will see, and they may see considerably less than we might hope. Likewise, as the commonplace (but erroneous) inference "$p_1p_2 \ldots p_n + 1$ is itself always prime" demonstrates, they may see rather more than we intend.

There is something to be learned from the student who commented as follows on the generic proof that every prime number has a primitive root:

> Although the explanation for $p = 19$ is clear and true, it doesn't necessarily follow that $p = 29$ has a primitive root. So I'd prefer to work through it [$p = 29$] before I was convinced.

Whereas most of the other students were able to isolate the invariants of the argument, to track the role of 19 in the single generic example given, maybe this particular student *does* need to "work through" the case $p = 29$ to sense the architecture of the proof, to begin to perceive that it is intended to be much more than a confirming instance. This student may need to retain caution with regard to empirical generalization while developing awareness of the possibility of structural generalization.

However, students may be lured into believing that the *form* of an example offers more by way of explanation than it actually does. For example, the students who seemed to believe that $\phi(pq) = (p - 1)(q - 1)$ was assured for all distinct primes p, q by the confirming case $p = 5$, $q = 7$ have seriously misunderstood the structural, explanatory force to be required of an example if it is to be deemed "generic" in the sense of Mason and Pimm (1984). In calculating $\phi(35)$ as a response to an exercise, despite arriving at the "correct answer" 24, they were perhaps insufficiently aware of the combinatorial process that led them to that conclusion. Their awareness of their own actions is perhaps in need of educative attention. They need to be helped to attend to such actions in a way that they have not until now. That, in turn, is an educational task that has been addressed in some depth in the mathematics education literature (for example, by Hewitt, 1994, 1997). Once we as teachers, in turn, are aware (and convinced) of the importance of this awareness in our students, we are in a better position to address it as a pedagogical issue.

Pedagogical Suggestions and Proposals for Further Research

I detect in the "mathematical community" a commonly held view that generic proofs are formally inadequate, being at best a staging post between naive empiricism and general argument. Even if we accept that position, it would be worth expending some effort in developing the status and the didactics of that staging post. In fact, I would now question that conventional view. Being persuaded by Hersh's argument that, in the classroom, the paramount purpose of proof is to explain, to illuminate, I believe that there are moments in the learning of number theory when well-constructed generic examples can be a complete and sufficient means of achieving that purpose. In the spirit of Semadeni (1984), I see no reason to comply with the demands of conventional symbolic formality priority over pedagogical responsibility to one's students. In my role as a teacher, my responsibility is quite the reverse. I believe that learners of mathematics at all levels, including university students, should be assisted to perceive and value that which is generic in their particular insights, explanations, and arguments.

But what about those formal proofs in general symbolic notation? Here, I have taken an extreme line in questioning their necessity. Let me conclude by taking a more moderate stance, prompted by the caution of a respected Cambridge mathematician.

> There is a sense in which you are reversing a familiar step. If a student does not understand a proof one often suggests that he/she "works it through in a particular case." What you are suggesting, I think, is that you reverse the order here with the added benefit that the student is confident before the formal proof rather than being depressed after it. . . . Most university lecturers would, I think, take the view that a significant part of the training is about teaching students to be able to write and understand formal proof, and not just the ideas behind the proof . . . I think that if you want to have a good chance of success you must not only develop the idea of the generic example, but also show how it really does lead on to a formal proof. To ignore the formal proof at the end will, I think, greatly diminish the acceptance of the idea. (A. Beardon, personal communication, February 7, 2000)

Having internalized the argument in a particular generic example, and having perceived its structural relationship to other possible examples, how can students be enabled to write "proper" proofs, to bridge the gap between generic understanding and general exposition? The transition between the first kind of knowing and the second seems to entail the harnessing of ideas to notation. My students' formalization of the proof of Wilson's Theorem shows that it may be accomplished in arguments of a relatively straightforward nature. However, Principle 5 acknowledges that this is no trivial matter in sustained and multilayered arguments, such as that presented in the primitive roots proof. For the moment, I would offer three modest and conservative suggestions for supporting students, summarizing the discussion of a number of points in this chapter.

- Choose and construct your generic examples with care, following Principles 1 to 4.
- Following presentation of the generic argument, invite students to say what it would "look like" in the case of another particular case, for example, another particular prime number.
- Finally, get students to write out the argument for the general case. Discuss such transitions with the whole class if you have the time. In any case, scrutinize students' written responses to try to infer what they discern in the example, and what help they seem to need with notation.

While one can cheerfully and optimistically speculate about tracking variables, about isolating relations between mathematical objects, we need more than dependence on the generic intuition of teachers like Kate and Alison. A professional task ahead for teachers of number theory and for researchers in mathematics education might be to consider the principles enunciated in this chapter, to try them out and to refine them. This would seem to offer a good prospect of developing and offering a systematic didactic technology of formal proof in number theory by building on skillfully constructed generic examples.

NOTES

1. Hersh's use of "conviction" is somewhat idiosyncratic. Most mathematics educators associate the word with absence of doubt about the truth of a proposition, with the same sense that Harel and Sowder (1998) use the term "ascertaining": see also Harel's contribution to this volume. This is also the sense in which Copes (cited later in this chapter) uses "convinced."
2. Given the time for which the world has waited for a proof of Fermat's Last Theorem, Wiles's particular demonstration might be deemed to be "recent" for some time to come.
3. In the case of this theorem, there is no shortage of proofs to choose from (Loomis, 1968).
4. *Teachers Count*. Ofsted, Crown Copyright, 1997.
5. This scenario delightfully pushes to the limit the notion of "real" problem solving, and challenges a preoccupation with "relevance." I like the Dutch notion "realistic," which, as I understand it, refers to contexts that engage and motivate learners. The world of verifiable experience can be a little limiting at times.
6. Such actions are characteristically of the same kind as the goal-oriented "operations on objects" involved in the transformational proof schemes of Harel and Sowder (1998).
7. Some readers may be more familiar with "relatively prime," which has the same meaning as "co-prime."
8. The exposition given here observes the spirit, rather than the letter, of Walther's argument. The modifications presented in this account are inspired by Burn (1982).
9. The simplest version of the Chinese Remainder Theorem asserts that if m_1 and m_2

are co-prime, the two congruences $x \equiv a \bmod m_1$, $x \equiv b \bmod m_2$, are simultaneously satisfied by an integer, which is unique modulo $m_1 m_2$. The entry in each cell of the matrix shown is the unique integer in the range 0 to 23, which is congruent with a mod 3 *and* b mod 8 ($0 \leq a < 3$, $0 \leq b < 8$).

10. Of course, this also needs justification. If t and 24 are not co-prime, then there exists a *prime* number p such that $p \mid t$ and $p \mid 24$. But $24 = 3 \times 8$ and $p \mid ab \Rightarrow p \mid a$ or $p \mid b$.

11. As will become apparent later, the theorem is a significant preliminary to the proof that every prime number has a primitive root.

12. If $1 \leq a \leq p - 1$, and $a^2 \equiv 1 \bmod p$ then $p \mid (a - 1)(a + 1)$ and so $p \mid (a - 1)$ or $p \mid (a + 1)$. Hence $a = 1$ or $a = p - 1$. The essence of this argument is the solution of a quadratic equation in a multiplicative modular group, which seems to rule out generic presentation.

13. The canonical *reduced* set of residues modulo n consist of those integers in the range 1 to $n - 1$, which are co-prime with n. When, as in the case of 13, n is prime, this reduced set of residues is $\{1, 2, 3, \ldots, n - 1\}$. "Reduced" is in contrast to the (least positive) *complete* set of residues $\{0, 1, 2, \ldots, n - 1\}$ for all integers n, whether prime or composite. The complete set contains one representative for each of the n congruence classes modulo n. The complete set and the reduced set of residues modulo n form, respectively, additive and multiplicative groups.

14. I am grateful to the editors for confirming what I might have suspected, that this observation is by no means original. See Leron (1985).

15. To the consternation of novice number theory students, a plethora of equivalent statements is possible, every one of which can be found in some textbook or another. I was obliged to pick one formulation.

16. I owe this suggestion to John Mason.

17. Only holders of QTS may teach in maintained ("state") schools in England. An alternative route to QTS is a first degree in a mainstream school subject, such as mathematics, and a one-year course of professional training leading to a Postgraduate Certificate in Education (PGCE).

18. This reasoning is on the basis of Lagrange's Theorem on subgroups, which states that the order of (i.e., the number of elements contained in) every subgroup divides the order of the group. In this case, the group is M_{19} with order 18, and the order of every element a is equal to the order of the subgroup generated by a. Note that "order" is being used here in two distinct, but closely related, senses.

19. Having been in my company for over thirty hours so far, they are familiar with the term "generic example" and the corresponding pedagogical intention.

20. Recall the notation M_p for the group $\{1, 2, 3, \ldots, p - 1\}$ under multiplication mod p.

21. "Knowing Why: Generic Examples as Didactic Tools." November 16, 1999, Department of Educational Studies, University of Oxford, UK.

REFERENCES

Baker, A. (1984). *A concise introduction to the theory of numbers*. Cambridge: Cambridge University Press.

Balacheff, N. (1988). Aspects of proof in pupils' practice of school mathematics. In D.

Pimm (Ed.), *Mathematics, teachers and children* (pp. 216–235). London: Hodder and Stoughton.

Bell, E. T. (1937). *Men of mathematics*. Harmondsworth, UK: Penguin.

Bills, L. (1996). The use of examples in the teaching and learning of mathematics. In L. Puig & A. Gutierrez (Eds.), *Proceedings of the 20th Conference of the International Group for the Psychology of Mathematics Education* (Vol. 2, pp. 81–88). Valencia, Spain: University of Valencia.

Burn, R. P. (1982). *A pathway into number theory*. Cambridge: Cambridge University Press.

Collis, K. F. (1972). *A study of concrete and formal operations in school mathematics*. Unpublished doctoral thesis, University of Newcastle, New South Wales.

Copes, L. (1980). Constructing castles. *Mathematics Teaching, 90*, 43–45.

Fauvel, J. (1987). *The Greek concept of proof*. Milton Keynes, UK: Open University Press.

Foxman, D., Ruddock, G., Badger, M., & Martini, R. (1982). *Mathematical development: Primary survey report no. 3*. London: Her Majesty's Stationary Office.

Hanna, G. (1989). Proofs that prove and proofs that explain. In G. Vergnaud, J. Rogalsky, & M. Artique (Eds.), *Proceedings of the Thirteenth International Conference for the Psychology of Mathematics Education* (Vol. 2, pp. 45–51). Paris, France.

Harel, G., & Sowder, L. (1998). Students' proof schemes: Results from exploratory studies. In A. H. Schoenfeld, J. Kaput, & E. Dubinsky (Eds.), *Research in collegiate mathematics education, III* (Vol. 7, pp. 234–283). Providence, RI: American Mathematical Society.

Hersh, R. (1993). Proving is convincing and explaining. *Educational Studies in Mathematics, 24*, 389–399.

Hewitt, D. (1994). *The principle of economy in the learning and teaching of mathematics*. Unpublished doctoral thesis, Open University Press, Milton Keynes, UK.

Hewitt, D. (1997). Teacher as amplifier, teacher as editor. In E. Pehkonen (Ed.), *Proceedings of the 21st Conference of the International Group for the Psychology of Mathematics* (Vol. 3, pp. 37–80). Helsinki, Finland: University of Helsinki.

Hoffman, P. (1998). *The man who loved only numbers*. London: Fourth Estate.

Leron, U. (1985). A direct approach to indirect proofs. *Educational Studies in Mathematics, 16*(3), 321–326.

Loomis, E. S. (1968). *The Pythagorean proposition*. Reston, VA: National Council of Teachers of Mathematics.

Mason, J., & Pimm, D. (1984). Generic examples: Seeing the general in the particular. *Educational Studies in Mathematics, 15*, 277–289.

Morley, A. (1967). Changes in primary school mathematics—are they complete? *Mathematics Teaching, 41*, 20–24.

Nelson, R. B. (1993). *Proofs without words: Exercises in visual thinking*. Washington, DC: The Mathematical Association of America.

Polya, G. (1962). *Mathematical discovery, Volume I*. New York: Wiley.

Rowland, T. (1998). Conviction, explanation and generic examples. In A. Olivier & K. Newstead (Eds.), *Proceedings of the 22nd Conference of the International Group for the Psychology of Mathematics Education* (Vol. 4, pp. 65–72). Stellenbosch, SA: University of Stellenbosch.

Rowland, T. (1999). *The pragmatics of mathematics education: Vagueness in mathematical discourse*. London: Falmer Press.

Semadeni, Z. (1984). Action proofs in primary mathematics teaching and in teacher training. *For the Learning of Mathematics, 4*(1), 32–34.

Singh, S. (1997). *Fermat's last theorem.* London: Fourth Estate Limited.

Walther, G. (1984). Action proof vs. illuminating examples? *For the Learning of Mathematics, 4*(3), 10–12.

Walther, G. (1986). Illuminating examples: An aspect of simplification in the teaching of mathematics. *International Journal of Mathematical Education in Science and Technology, 17*(3), 263–273.

10

The Development of Mathematical Induction as a Proof Scheme: A Model for DNR-Based Instruction[1,2,3]

Guershon Harel

Undergraduate programs in mathematics usually do not include a course in number theory, but students are introduced to elementary number theory concepts in other courses, most often in a discrete mathematics course, which usually includes the concepts of mathematical induction (MI). MI is a prominent proof technique in discrete mathematics and number theory, where it is used to prove theorems involving properties of the set of natural numbers. Fermat, the founder of number theory, used a form of mathematical induction to prove many of his discoveries in this field (Boyer, 1968). Beyond its significance as a proof technique in mathematics, MI can provide a context to enhance students' conception of proof, as we see later in this chapter.

"There are unresolved problems concerning the teaching of MI which should benefit from a careful analysis" asserts Ernest (1984, p. 173) and adds that "there is . . . no systematic account in print of the teaching of MI, of the problems that arise, of the deeper issues involved or of the treatments given by textbooks" (p. 174). With the exception of Dubinsky's (1986, 1989) work, little has been done in research on the learning and teaching of MI during the last two decades. One goal of this chapter is to revive interest in this area.

The research reported here is part of the PUPA project.[4] The general questions addressed in PUPA revolved around the development of college students' *p*roof *u*nderstanding, *p*roduction, and *a*ppreciation: What are students' (particularly mathematics major students') conceptions of proof? What sorts of experiences seem effective in shaping students' conception of proof? Are there promising frameworks for teaching the concept of proof so that students appreciate the value of justifying, the role of proof as a convincing argument, the need for rigor, and the possible insights gained from proof? In answering these questions PUPA produced two complementary products: (a) a conceptual framework for students' proof schemes and (b) a system of pedagogical principles, called DNR—an acronym for the three leading principles in the system: *duality, necessity*, and *repeated-reasoning*. The two products complement each other in that while the former aimed at mapping students' current conceptions of proof, the latter specifies foundational pedagogical principles for enhancing students' proof schemes.

The investigation into students' conception of MI reported in this chapter is an example of a realization of these two complementary products. The first section of the chapter presents an overview of the conceptual framework for students' proof schemes (a complete description of the framework is in Harel & Sowder, 1998, and Harel, in press). The second and third sections analyze students' conceptions of MI in terms of these proof schemes; the former focuses on those recorded in standard instruction, whereas the latter on those developed in an alternative, DNR-based instruction. The fourth, and last, section describes briefly the DNR system, with particular reference to the alternative instructional treatment of MI as an example of its implementation. A full description of the DNR system appears in Harel (in preparation).

THEORETICAL PERSPECTIVE

The tools and language of the analyses presented in this chapter are Harel and Sowder's (1998) conceptual framework for students' proof schemes. The framework has recently been revised to reflect new observations, both theoretical and empirical (Harel, in press). Since it is not essential, nor possible due to space limitations, to describe here the entire framework, only those aspects of the framework that are relevant to the concerns of this chapter will, as needed, be briefly described. An overall view of the framework is worth depicting, however. Figure 10.1 gives a bird's-eye view of the framework structure. As can be seen, the framework consists of three classes of proof schemes: the *external conviction* proof scheme class, the *empirical* proof scheme class, and the *deductive* proof scheme class (to be defined below).

Critical to this conceptual framework is the definition of *proof scheme*[5] (given in Harel & Sowder, 1998, pp. 241–244). *Proving* is defined there as "the process employed by a person to remove or create doubts about the truth of an observation" and a distinction is made between two processes of proving: *ascertaining* and *persuading*. "Ascertaining is a process an individual employs to remove her or his own doubts about the truth of an observation. . . . Persuading is a process an individual employs to remove others' doubts about the truth of an observation." Thus, "a person's proof scheme consists of what constitutes ascertaining and persuading for that person."

Figure 10.1.
A conceptual framework for students' proof schemes consisting of three classes of proof schemes. The arrow from the transformational proof scheme category to the modern proof scheme category indicates that transformational reasoning is always present in the modern axiomatic proof reasoning, but the converse is not necessarily true.

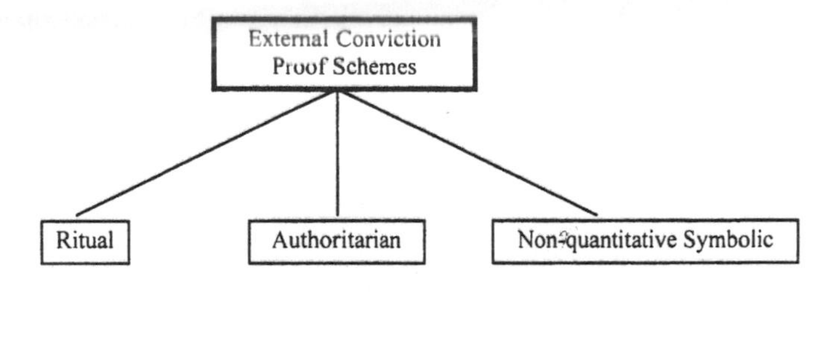

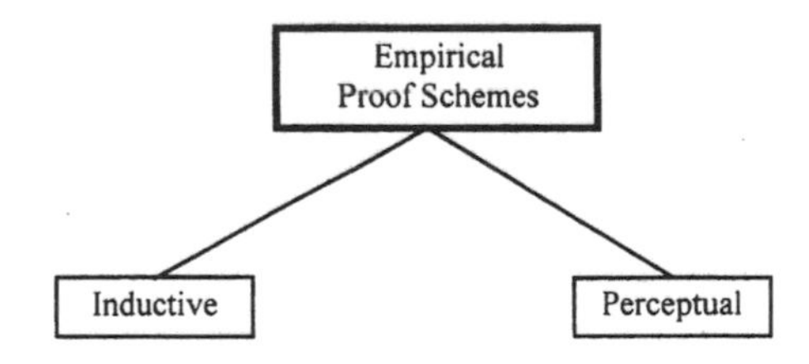

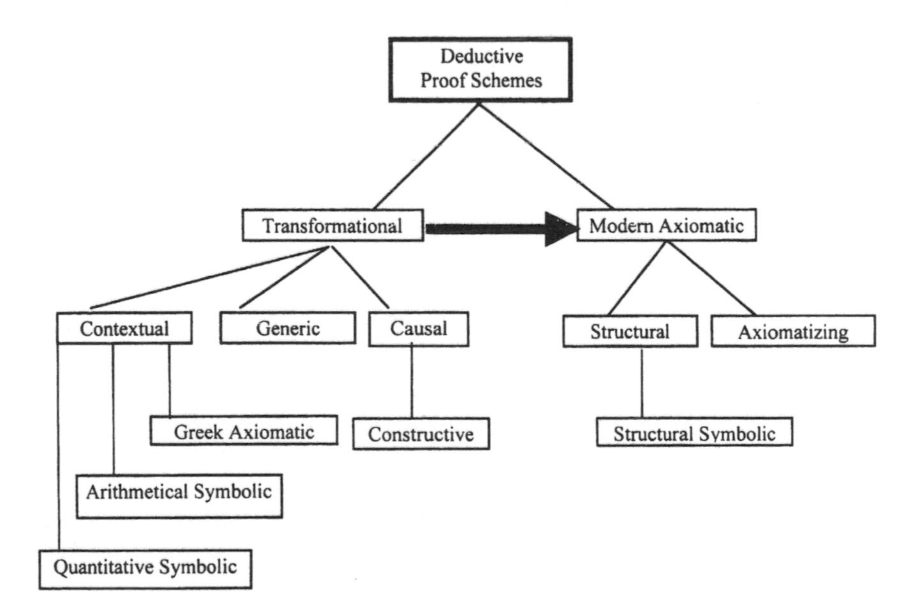

Seldom do these processes occur in separation: in ascertaining oneself, one considers how to persuade others, and vice versa. Thus, proof schemes are subjective and vary from person to person, civilization to civilization, and generation to generation within the same civilization. Yet the goal of instruction must be unambiguous; namely, to gradually refine current students' proof schemes toward the proof scheme shared and practiced by the mathematicians of today. This claim is based on the premise that such a shared scheme exists and is part of the ground for scientific advances in mathematics.

Research Questions and Method

Surprisingly, published studies on the learning and teaching of MI have not linked students' behaviors with MI to their conception of proof. Dubinsky's (1986, 1989) work includes proof as a factor in his genetic decomposition of MI, but he neither characterizes students' proof schemes nor addresses their specific role in the learning process of MI. A major goal of this chapter is to fill this gap by answering two questions:

1. How are students' conceptions of and difficulties with MI in a standard instructional treatment a manifestation of their proof schemes?
2. What are students' difficulties with and conceptual development of MI in an alternative instructional treatment?

The alternative treatment was implemented in a teaching experiment in elementary number theory (hereafter, the NT experiment). Twenty-five junior prospective secondary teachers took part in the NT experiment. It included the concepts of divisibility, greatest common divisors, modular arithmetic, and Diophantic equations; algorithms from number theory, such as the algorithms for finding the greatest common divisor of two positive integers; applications to computer arithmetic; cryptology; and MI. The goal of the experiment was to record students' proof schemes and the progress of their development. This chapter, however, focuses on a two-week period of the NT experiment during which a fundamentally different instructional treatment of MI was implemented.

The first question is answered on the basis of observations from several sources (to be referenced below), including our own, which have recently been analyzed and interpreted in terms of the proof scheme framework (Figure 10.1). The second question, on the other hand, was a primary research target of the NT experiment—one in a sequence of six teaching experiments on the development of college students' concept of proof.

"Teaching experiment" here is as defined by Cobb and Steffe (1983): Each teaching session is analyzed in terms of the classroom discourse and students' performance. The result of the analysis can, and usually does, adjust or amend

Figure 10.2.
A domino analogy to MI.

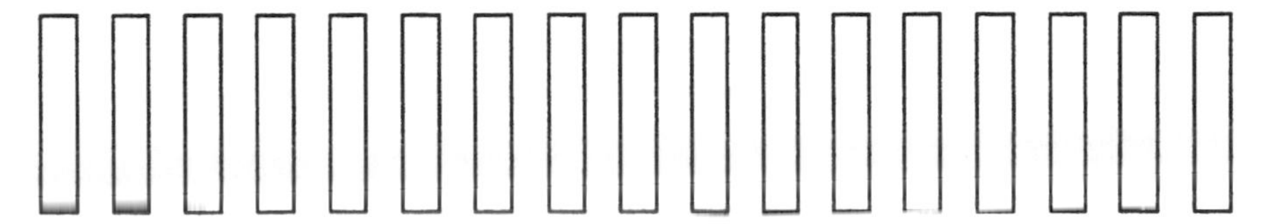

the plan for subsequent lessons. Results accumulated from extensive analyses usually refine, and in some cases alter, the researchers' theoretical perspective.

In our teaching experiments, the data consist of classroom observations in the form of field notes and retrospective notes, clinical interviews, homework, written tests and quizzes, videotaped classroom sessions, and clinical interviews with students.

STUDENTS' PROOF SCHEMES OF MI IN STANDARD INSTRUCTIONAL TREATMENT

A typical instructional approach to MI, as it is presented in standard undergraduate mathematics textbooks, begins with a few examples of how formulas with a single variable of a positive integer n (e.g., the formula for the sum of the first n positive odd integers) is generalized from an observed pattern. Following this, the question of how to prove propositions that state $P(n)$ is true for all positive integers n is discussed and answered by stating the principle of MI as a proof technique consisting of two steps:

1. *Basic step.* Show $P(1)$
2. *Inductive step.* Show $P(n) \Rightarrow P(n + 1)$ for every positive integer n.

In discussing this technique some textbooks alert the reader to the common misconception that a proof by MI is a case of circular reasoning, and stress that the proof does not assume that $P(n)$ is true for all positive integers, a common confusion among students.

In an attempt to help students understand how the principle works, some textbooks present the domino analogy to MI: The teacher points to a figure of an infinite row of dominos, labeled 1,2,3, . . . ,n, . . . (like in Figure 10.2) and asks the students to consider the proposition $P(n)$ stating domino n is knocked over. The teacher then explains that if the first domino is knocked over—that is, if $P(1)$ is true—and if whenever the nth domino is knocked over, it also knocks the $(n + 1)$st domino—that is if $P(n) \Rightarrow P(n + 1)$ is true—then all the dominos are knocked over.[6]

The quality of a concept image (Vinner, 1985) is impacted not only by how the concept is presented but also by how the concept is applied in solving problems. It is important, therefore, to also examine the kind and order of problems textbooks assign students to practice the principle of MI. For this, the following classification is needed.

MI problems in standard textbooks can be classified into two categories, *recursion problems* and *nonrecursion problems*, according to whether the mathematical solution of the problem requires the formation of a recursive representation of a function. For example, the problem, "Prove that $1 + 3 + 5 + \ldots + (2n - 1) = n^2$ for all positive integers n," is a recursion problem, because the left-hand side of the identity must be interpreted as a recursive representation of a function.[7] In contrast, the problems, "Prove that $n < 2^n$ for all positive integers $n > 3$" does not involve a recursive representation of a function, and so it is a nonrecursion problem. In some cases the recursive representation of a function is explicit in the problem statement, in others implicit. For example, in the former problem the recursive representation of a function is explicit (the rule, $f(n) = f(n - 1) + (2n - 1)$, $f(1) = 1$, is virtually present in the problem statement), whereas in the Tower of Hanoi problem or in the problem, "Find an upper bound to the sequence $\sqrt{2}, \sqrt{2+\sqrt{2}}\ \sqrt{2+\sqrt{2+\sqrt{2}}}, \ldots$ is implicit, for no recursively defined function is explicitly present in the problem statement. Accordingly, recursion problems are further classified into two categories: *explicit recursion problems* and *implicit recursion problems* (Figure 10.3). It should be noted, however, that the solver might not know what a recursive representation of a function is or that such a representation is involved in the problem.

Typically in the standard instructional approach to MI, students' first exposure to MI is through three kinds of problems: (a) identity problems (e.g., "Prove that $1 + 3 + 5 + \ldots + (2n - 1) = n^2$ for all positive integers n); (b) inequality problems (e.g., "Prove that $n < 2^n$ for all positive integers $n > 3$); and (c) divisibility problems ("Prove that $3 \mid n^3 - n$ for all positive integers n"). Implicit recursion problems, such as the Tower of Hanoi Problem, in which one is required to form a recursive representation of a function, usually are not included or appear in a small number at the end of the exercises list. As will be argued later in this chapter, the type and sequence of MI problems is of paramount pedagogical importance.

I began looking closely into the learning and teaching of MI about 10 years ago. Until then my instructional treatment followed the standard approach—both in content presentation and in the kind and order of problems assigned. My students' understanding of MI was unsatisfactory. Their difficulties were consistent with those reported in the literature, and in retrospect, when analyzed in terms of students' conceptions of proof, they were found to be mostly manifestations of the *external* and *empirical* proof schemes (Figure 10.1), as we will now see.

Figure 10.3.
A classification of problems involving proposition-valued functions whose domain is the natural numbers.

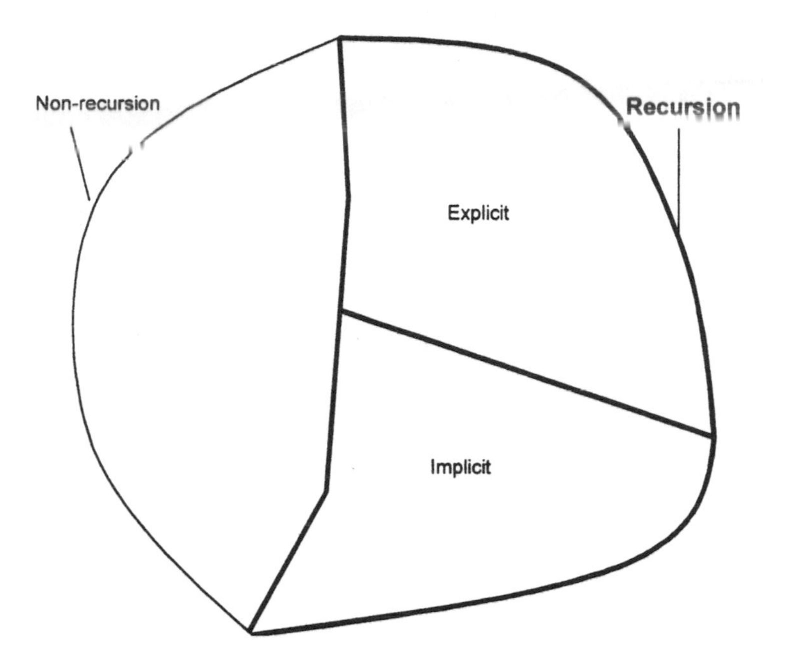

MI as an Empirical Proof Scheme: The Case of Result Pattern Generalization

One of the difficulties students have with MI, also observed by Show (1978) and reviewed in Ernest (1984), is their view of induction as a technique of drawing a general conclusion from a number of individual cases. This observation, however, clomps two distinct ways of thinking into one: in one, students generalize from a *process pattern*, in the other from *result pattern*. In *process pattern generalization* students focus on regularity in the process, whereas in *result pattern generalization* on regularity in the result. For example, some students responded to the log-identity problem (Prove that for all positive integers n, $\log(a_1 \cdot a_2 \ldots . a_1) = \log a_t + \log a_2 + \ldots + \log a_t$) by relying on the regularity of the results of their calculations, as in the following response by a student:

Response 1:
$\log(4 \cdot 3 \cdot 7) = \log 84 = 1.924$ $\quad$ $\log(4 \cdot 3 \cdot 6) = \log 72 = 1.857$
$\log 4 + \log 3 + \log 7 = 1.924$ $\quad$ $\log 4 + \log 3 + \log 6 = 1.857$
Since these work, then $\log(a_1 \cdot a_2 \cdot \ldots \cdot a_n) = \log a_1 + \log a_2 + \cdots + \log a_n$.

A probe into these students' reasoning has revealed that their conviction stems from the fact that the proposition is shown to be true in a few instances, each with numbers that are *randomly* chosen. This behavior is a manifestation of the empirical proof scheme, a scheme in which students rely on evidence from perception or examples of direct measurements of quantities, substitutions of specific numbers in algebraic expressions, and so on (Harel & Sowder, 1998).

MI as a Deductive Proof Scheme: The Case of Process Pattern Generalization

Process pattern generalization is a way of thinking in which one's conviction is based on regularity in the process, though it might be initiated by regularity in the result. This behavior is in contrast to the result pattern generalization, where students' proving is based solely on regularity in the result—obtained by substitution of numbers, for instance.

To demonstrate the process pattern generalization, consider the following response by one of the students to the log-identity problem we just discussed:

> *Response 2*:
> (1) $\log(a_1a_2) = \log a_1 + \log a_2$ by definition
> (2) $\log(a_1a_2a_3) = \log a_1 + \log a_1a_2$. Similar to $\log(ax)$ as in step (1), where this time $ax = a_1a_2$.
> Then
> $\log(a_1a_2a_3) = \log a_1 + \log a_2 + \log a_3$
> (3) We can see from step (2) any $\log(a_1a_2a_3 \cdots a_n)$ can be repeatedly broken down to $\log a_1 + \log a_2 + \cdots + \log a_n$

In this excerpt the student recognizes that the process he employs in the first and second cases constitutes a pattern that recursively applies to the entire sequence of propositions, $\{\log(a_1a_2a_3 \cdots a_n) = \log a_1 + \log a_2 + \cdots + \log a_n\}_n$.

Both the result pattern generalization and the process pattern generalization are based on a limited number of specific cases—in both Responses 1 and 2 above, for example, the generalizations are made from two cases. This may suggest, therefore, that both kinds of generalizations are empirical. This is not so, however. The process pattern generalization is *not* a manifestation of the empirical proof scheme but an expression of the transformational proof scheme. To see why, one only needs to examine the two generalizations against the definitions of the two schemes. The transformational proof scheme is characterized by (a) consideration of the generality aspects of the conjecture, (b) application of mental operations that are goal oriented and anticipatory—an attempt to predict outcomes on the basis of general principles—and (c) transformations of images that govern the deduction in the evidencing process (Harel & Sowder, 1998).[8] While both responses share the first characteristics—that is, in both the students' responses to the "for all" condition in the log-identity problem statement—they differ in the latter two: Whereas the mental operations in Response 1 are incapable of anticipating possible subsequent outcomes in the sequence and are devoid of general principles in the evidencing process, the mental operations in Response 2 correctly predict, on the basis of the general rule $\log(ax) = \log a + \log x$, that the same outcome will be obtained in each step of the sequence. Further, in Response 1 the inference rule that governs the evidencing process is empirical; namely, $(\exists\, r \in R)(P(r)) \Rightarrow (\forall\, r \in R)(P(r))$. In Response 2, on the other hand, it is deductive; namely, it is based on the inference rule $(\forall\, r \in R)(P(r)) \wedge (w \in R) \Rightarrow P(w)$. (Here, r is any pair of real numbers a and x; R is the set of all pairs of real numbers; $P(r)$ is the statement "$\log(ax) = \log a + \log x$" and w in step n is the pair of real numbers, $a_1a_2 \ldots a_{n-1}$ and a_n.)

MI as an External Proof Scheme

Dubinsky (1986) has indicated "if you question students—even those who have had several mathematics courses—although almost all of them will have heard of induction, not many of them will be able to say anything intelligent about what it is, much less actually use it to solve a problem" (p. 305). Our experience corroborates Dubinsky's observation. "It is a proof with steps" was the only response to the instructor's question "What is MI?" despite the fact that the entire class confirmed that they were familiar with the term MI.

When pressed, some students admit that they view the step of verifying P(1) as nonessential, and one is required to perform it just to satisfy the rule stated by the teacher. For, according to these students, the reason that $P(n)$ is true for all positive integers n is that: "We proved it for $n + 1$, so we proved it for n." Seldom do students express doubts about the process of MI; they accept it, as they have been trained to accept without question any other mathematical statement. This behavior is a manifestation of the authoritative proof scheme (Figure 10.1)—a scheme in which proving depends mainly on the authority of the teacher or textbook.

In the authoritative proof scheme, one follows the rule prescribed by the principle of MI without understanding its meaning: first verify P(1), then follow the procedure of assuming $P(n)$, go through a certain manipulation, and then derive $P(n + 1)$ (Baxandall, Brown, Rose, & Watson, 1978); thus reinforcing another external proof scheme, the *symbolic nonquantitative* proof scheme—a scheme by which one thinks of symbols as if they possess a life of their own without reference to their functional or quantitative meaning (Harel & Sowder, 1998). As was observed by Woodall (1981, p. 100), for many students, induction means, "Take an equation involving n and add something to both sides so as to produce a similar equation with $n + 1$ in place of n" (quoted in Ernest, 1984). Many students perform this procedure correctly without understanding what they are doing. For example, some students did not know what the "something" they are adding to both sides of the equality was, as the following case illustrates:

> A student was given the problem: "Prove that for any positive integer n, $\frac{1}{1 \cdot 2} + \frac{1}{2 \cdot 3} + \cdots + \frac{1}{n \cdot (n+1)} = \frac{n}{n+1}$." He first performed the *basic step*, for $n = 1$, $\frac{1}{1 \cdot 2} = \frac{1}{2}$, and then added $\frac{1}{(n+1)(n+2)}$ to both sides of the equality, but did not know whether he was adding the $(n+1)^{st}$ term of the sequence or the sum of the first $n + 1$ terms. Although he performed the algebraic computation correctly and obtained the expression $\frac{n+1}{n+2}$ as he desired, his understanding of the process was merely procedural, not relational.

In summary, in the standard instructional treatment of MI the three most prevalent proof schemes among students are the authoritative proof scheme; the symbolic, nonquantitative proof scheme; and the result pattern generalization, which is a manifestation of the empirical proof scheme. On the other hand, process pattern generalization—an expression of the transformational proof scheme—is uncommon among students. Its mere occurrence, however, offers an important hint as to an alternative, more effective instructional approach to MI, which is the topic of the next section.

STUDENTS' PROOF SCHEMES OF MI IN AN ALTERNATIVE INSTRUCTIONAL TREATMENT

The standard instructional treatment of MI has two major deficiencies. The first deficiency is that the principle of MI is introduced abruptly: students do not see how the principle is born out of a *need* to solve problems, nor do they see how it is derived from previous, more elementary experience. Rather, it is handed to them as a prescription to follow; thus reinforcing the authoritative and symbolic, nonquantitative proof schemes.

The second deficiency is in the type and order of problems used. The significance of implicit recursion problems, as will be demonstrated later in this section, is that they induce students to focus on process pattern generalization rather than on result pattern generalization, whereby they build and reinforce the transformational proof scheme. As has been indicated earlier in this chapter, standard textbooks focus almost exclusively on identity problems, inequality problems, and divisibility problems. These problems require little understanding of MI, for students learn to solve them by blindly following the two MI steps: verify P(1) and then go through a certain manipulation to obtain P(n + 1)—not necessarily understanding that the latter is derived from P(n).

Some textbook authors, in an attempt to sequence the problems assigned to students in what they perceive as an increasing order of difficulty, begin with "easy" problems, such as "Prove by MI the statement, for any positive integer $n \geq 4$, $n! > 2^n$," and cause as a result an unexpected hindrance (see below the subsection, "Essentiality of MI"). Most of the students in our teaching experiments viewed such statements as "trivial," as ones that do not require MI but can be solved in an "easier" way. The demand to use MI for these statements was, therefore, nonintrinsic, alien, to the students. This is not to say, however, that such problems should not be assigned. As we will see below, they can play an important role in the interiorization[9] process of MI, provided they are introduced in a suitable stage in the students' conceptual development.

The new instructional treatment of MI takes into considerations these deficiencies. It is structured to help students develop the principle of MI gradually through intrinsic problems—problems they can understand and appreciate. The treatment consists of three phases corresponding to the three levels of conceptual development:

1. Quasi-induction as an internalized process pattern generalization,
2. Quasi-induction as an interiorized process pattern generalization, and
3. MI as an abstraction of quasi-induction.

These phases will be discussed in this order in the next three subsections.

Phase 1: Quasi-induction as an Internalized Process Pattern Generalization

Students' first exposure to MI was through engagement with implicit recursion problems typified by:

1. Find an upper bound to the sequence $\sqrt{2}, \sqrt{2+\sqrt{2}}\ \sqrt{2+\sqrt{2+\sqrt{2}}}, \ldots$
2. You are given 3^n coins, all identical except for one, which is heavier. Using a balance, prove that you can find the heavy coin in n weighings.
3. Let n be a positive integer. Show that any $2^n \times 2^n$ chessboard with one square removed can be tiled using L-shaped pieces, where these pieces cover three squares at a time.
4. *The Tower of Hanoi Problem*: Three pegs are stuck in a board. On one of these pegs is a pile of disks graduated in size, the smallest being on top. The object of this puzzle is to transfer the pile to one of the other two pegs by moving the disks one at a time from one peg to another in such a way that a disk is never placed on top of a smaller disk. How many moves are needed to transfer a pile of n disks?
5. Prove that every third Fibonacci number[10] is even.
6. Into how many regions is the plane cut by n lines, assuming no two lines are parallel and no three intersect at a point.

In the standard instructional approach, these problems are usually kept to the end of the exercise list or not introduced at all. The students in the NT teaching experiment, on the other hand, were engaged for a relatively long period of time—before the principle of MI was mentioned—in working on problems of this kind. These problems proved to be effective in shifting students' attention from result pattern generalization to process pattern generalization. For example, in working on Problem 1 in small groups, students first noticed that 2 is an upper bound—typically by finding on their calculators the values of several items in the sequence. Some students saw this as sufficient evidence—a typical result pattern generalization phenomenon. Other students established a process pattern generalization, as did one of the small working groups, who argued via their representative that: "The third item is less than 2 because it is the square root of a number that is smaller than 4, this number being the sum of 2 and a number that is smaller than 2." The representative proceeded by applying this form of argument to the next few terms in the sequence and concluded that all the terms in the sequence must be less than 2 because, according to her, the same relationship exists between any two consecutive terms in the sequence. While the result pattern generalization argument persuaded only some of the students, this process pattern generalization argument convinced all the students; hence, the class as a whole seemed to view the latter as superior to the former.

The language and notation used by the students within the class discussion developed gradually. For example, during the discussion of Problem 1 the students hardly used any mathematical symbols in referring to the terms in the sequence. In the discussion of Problem 4, on the other hand, they used mathematical symbols to convey their solution, as the following response by one of the small working groups illustrates:

> [The] formula is derived for moving the disks on 3 pegs. The formula is $2(S_n) + 1$. [Let's say] we moved n disks, and it took S_n moves. In order to move $n + 1$ disks, first we move n disks. This takes say S_n moves. Now, we move the $(n + 1)^{st}$ disk—that is the $+1$ of the formula. Last we move the n disks on top of the $(n + 1)^{st}$ disk, [which take another S_n moves]. We have now made a total of $2(S_n) + 1$ moves.

These students seem to understand that with the recursive formula, $S_{n+1} = 2S_n + 1$, one can compute the number of moves needed for any number of disks by simply starting from the number of moves needed for one disk and substitute it in the formula to find the number of moves needed for two disks, and so on.

Not all students pursued a process pattern generalization for these problems. Some students obtained a formula for the number of moves by result pattern generalization. For example, the representative of another working group presented the following table:

Disk #	# of moves
1	$1 = 2^1 - 1$
2	$3 = 2^2 - 1$
3	$7 = 2^3 - 1$
4	$15 = 2^4 - 1$
5	$31 = 2^5 - 1$

and then said: "Therefore, we can infer that $2^{31} - 1 = 2{,}147{,}483{,}647$, where $2^n - 1 = \#$ of moves and n = number of disks moved."[11]

This result intrigued some of the students. They saw that the two formulas constructed by the two working groups, $S_{n+1} = 2S_n + 1$ and $S_n = 2^n - 1$, produced the same outcome for any specific natural number n they chose. But this form of equivalency was foreign to them, because they did not know how to transform one formula into another—in the same way they knew how to transform trigonometric identities, for example. With direction from the instructor, the students easily established the implication $S_n = 2^n - 1 \Rightarrow S_{n+1} = 2S_n + 1$, but the converse was postponed to the third phase.

The current phase continued for about a week, during which the students repeatedly applied process pattern generalization in solving about twenty problems like Problems 1–6. The general solution approach that emerged from this *repeated reasoning* I refer to as *quasi-induction.* Students' expressions of quasi-induction are either in the form of the *method of ascent* or in the form of the *method of descent.*

The two methods will be exemplified now even though the latter appeared only in the second phase.

The method of ascent.

Problem: "Find an upper bound to the sequence $\sqrt{2}, \sqrt{2+\sqrt{2}}, \sqrt{2+\sqrt{2+\sqrt{2}}}, \ldots$ "

Student 1: Since $\sqrt{2}$ is less than 2, $2 + \sqrt{2}$ is less than 4, and so $\sqrt{2+\sqrt{2}}$ is less than 2. Since $\sqrt{2+\sqrt{2}}$ is less than 2, $2 + \sqrt{2+\sqrt{2}}$ is less than 4. Hence, $\sqrt{2+\sqrt{2+\sqrt{2}}}$ is less than 2. And so on.

Student 2: Each item of the sequence can be obtained from the item on [its] left in the same way [pointing to how she derived P(2) from P(1) and P(3) from P(2)], and this way is adding to the one on the left and taking the square root [i.e., $a_n = \sqrt{2 + a_{n-1}}$]

The method of descent.

Problem: Prove that $\frac{1}{1 \cdot 2} + \frac{1}{2 \cdot 3} + \frac{1}{3 \cdot 4} + \cdots + \frac{1}{n \cdot (n+1)} = \frac{n}{n+1}$ for all positive integers n.

Student 3:

$n = 2 \quad \frac{1}{2} + \frac{1}{6} = \frac{2}{3}$

$n = 3 \quad \frac{2}{3} + \frac{1}{2} = \frac{3}{4}$

$n = 4 \quad \frac{3}{4} + \frac{1}{20} = \frac{4}{5}$

$n = 5 \quad \frac{4}{5} + \frac{1}{30} = \frac{5}{6}$

$$\text{If } \frac{1}{1 \cdot 2} + \frac{1}{2 \cdot 3} + \frac{1}{3 \cdot 4} + \cdots + \frac{1}{n \cdot (n + 1)} = \frac{n}{n + 1}$$

$$\text{Then } \frac{1}{1 \cdot 2} + \frac{1}{2 \cdot 3} + \frac{1}{3 \cdot 4} + \cdots + \frac{1}{n \cdot (n + 1)} - \frac{1}{n \cdot (n + 1)}$$

$$= \frac{n}{n + 1} - \frac{1}{n \cdot (n + 1)}$$

$\Rightarrow$

$$\frac{1}{1 \cdot 2} + \frac{1}{2 \cdot 3} + \frac{1}{3 \cdot 4} + \cdots + \frac{1}{(n - 1) \cdot n} = \frac{n^2 - 1}{n(n + 1)}$$

$$= \frac{(n + 1)(n - 1}{n(n + 1)}$$

$$= \frac{n - 1}{n}$$

$$= \frac{n - 1}{[(n + 1) - 1]}$$

which is of the form $\frac{n'}{(n' + 1)}$

where $n = n - 1$.

These steps could be continued down to eventually get n to a small enough number so it can be evaluated easily. A few of these are shown above.

Problem: Prove that $\log(a_1 \cdot a_2 \cdot \ldots \cdot a_n) = \log a_1 + \log a_2 + \ldots + \log a_n$, for any positive integer n.

Student 4: Since $\log a_1 a_2 = \log a_1 + \log a_2$, if we look at $\log((a_1 \cdot a_2 \cdot \ldots \cdot a_{n-1}) a_n)$, we need to show $\log(a_1 \cdot a_2 \cdot \ldots \cdot a_{n-1}) = \log a_1 + \log a_2 + \ldots + \log a_{n-1}$. Then this equation can be reduced by the same method, until we get $\log a_1 a_2 = \log a_1 + \log a_2$.

The NT students invented quasi-induction while attempting to solve implicit recursion problems. They practiced the method by repeatedly applying it in solving problems similar to Problems 1–6 above. A total of twenty implicit recursion problems were assigned in two sets of homework assignments—eleven problems in the first set, nine problems in the second. After completing the first set of homework, students' application of quasi-induction, as was expressed in their individual work and the class discussions, seemed autonomous and spontaneous.

Quasi-induction, as the NT students conceived it in this instructional phase, may be viewed as a *theorem-in-action*, a term coined by Vergnaud (1988) to refer to "mathematical relationships that are taken into account by students when they choose an operation or a sequence of operations to solve a problem" (p. 144). A theorem-in-action is usually implicit knowledge, and so was quasi-induction in this instructional phase. More important, at this phase quasi-induction was an *internalized* theorem-in-action; namely, the students were able to apply it autonomously, and often spontaneously. The instructional goal, however, was that quasi-induction become an *interiorized* theorem-in-action; namely, that the students be able not only to apply quasi-induction autonomously, but also represent it conceptually as a method of proof—in the same way they might evoke proof-by-contradiction, for example.

Phase 2: Quasi-induction as an Interiorized Process Pattern Generalization

Implicit recursion problems used in the first phase necessitated for the students the formation of recursive relations, which in turn led them to construct, and with repeated application internalize, quasi-induction. Explicit recursion problems, such as Problems 7–9 below, were kept to the second phase, whose ultimate goal was to advance students to the state of conceiving quasi-induction as a method of proof.

7. Prove that for any positive integer $n \geq 1$,
$$\frac{1}{1 \cdot 2} + \frac{1}{2 \cdot 3} + \cdots + \frac{1}{n \cdot (n+1)} = \frac{n}{n+1}.$$
8. Find a formula for the sum $a + a^2 + a^3 + \ldots + a^n$
9. Compute the sum $1 + 3 + \ldots + 2n - 1$

The judgment to reserve explicit recursion problems to the second phase was this: The essential common feature to explicit recursion problems and implicit recursion problems is that the problem statement is about a proposition-valued function whose domain is the set of positive integers. Based on pilot experiments, we expected NT students not to recognize this fact spontaneously. But, we hypothesized, when they do, they would apply, or attempt to apply, quasi-induction to the new set of problems. And through this activity, the students would ascertain themselves and persuade others of the quasi-induction's applicability range expansion, whereby they would reflect on quasi-induction and recognize it as a method of proof.

This predicted scenario was in the main realized. At first, when explicit recursion problems were introduced, the students did not see any relation between them and the implicit recursion problems from the previous phase. Their approach to Problem 7, for example, was to find a closed form to the summation on the left-hand side of the equality—mostly by trying to find a common denominator to the fractional expressions. For Problem 8, some stated they knew the formula from their statistics class, but admitted they had never understood why it works. Interestingly, only a few students attempted to verify the equality in Problem 7 with specific positive integer values—an indication that most of the students did not see the problem statement as representing an infinite sequence of statements on the positive integers. Pointing out this fact to the students did not lead them to realize the common structure to the two sets of problems. The students had to be explicitly asked to see if such a common structure existed. Even when they recognized the common structure, they did not spontaneously attempt to apply quasi-induction to the new problems. Some of the students even fell back to mere empirical reasoning, especially in problems in which a result pattern generalization was apparent, such as in Problem 9. Here, many students observed easily that

$$1 = 1 = 1^2$$
$$1 + 3 = 4 = 2^2$$
$$1 + 3 + 5 = 9 = 3^2$$
$$1 + 3 + 5 + 7 = 16 = 4^2$$

and from this sequence, they concluded that $1 + 3 + \ldots + 2n - 1 = n^2$ for all positive integers n.

Only one of the small working groups discovered how to apply quasi-induction to solve the new problems. In expressing their solution to Problem 9, they explicitly referred to their solution to Problem 1 from Phase 1. The group's representative first represented the expression $1 + 3 + \ldots + 2n - 1$ as a sequence $1, 1 + 3, 1 + 3 + 5, \ldots$ and then said:

> Like in Problem 1, where the relationship between two consecutive elements a_k and a_{k+1} is $a_{k+1} = \sqrt{2 + a_k}$, here the relationship is $a_{k+1} = a_k + 2k + 1$. And like in Problem 1, where we used the fact that a_1 is smaller than 2 to derive that a_2 is smaller than 2, then used the fact that a_2 is smaller than 2 to derive that a_3 is smaller than 2, and so on, here we use the fact that $a_1 = 1^2$ to derive that $a_2 = 2^2$, because $a_2 = a_1 + 2 \cdot 1 + 1 = 1^2 + 2 \cdot 1 + 1 = 1^2 + 2 \cdot 1 + 1^2 = (1 + 1)^2 = 2^2$, and in a similar way to use that $a_2 = 2^2$ to derive that $a_3 = 3^2$, and so on.

This discovery was novel but not alien to the rest of the class—it was in the their zone of proximal knowledge. From this point on, the rest of the working groups began to adopt this pattern of reasoning in solving explicit recursion problems. The students continued for some time, especially in addressing their classmates, to point to parallels between their solutions to explicit recursion problems and specific solutions they used to solve implicit recursion problems—just as in the last excerpt.

Students' repeated attempts to attend to the underlying common structure of the two categories of problems—the implicit recursion problems from Phase 1 and the explicit recursion problems from Phase 2—and to apply a process pattern generalization approach in the form of quasi-induction to problems in the latter category led the students to reflect on the proof method they had acquired and successfully used in the previous phase. This together with the didactical contract established in class—that persuasion, not only ascertainment, is an essential part of proving—induced students to be explicit about their use of quasi-induction.

It is both interesting and important to note that historically quasi-induction was an antecedent to the concept of MI—a topic to be addressed in a sequel to this chapter (Harel, Manaster, McClure, & Sowder, in preparation). The institutionalization of quasi-induction as a means for persuasion and ascertainment was *not* achieved by mere democratic consensus, nor was it independent of the instructor's endorsement. It was from the very start an explicit instructional objective, for it was judged to correspond to and be the cognitive root (in the sense of Tall, 1992) for the principle of MI—a judgment supported by an analysis of the historical epistemology of MI. However, the process by which the students developed this method of proof was *largely* adidactical (Brousseau, 1997)—free from considerations of pleasing the teacher or conforming to his will.

Phase 3: MI as an Abstraction of Quasi-induction

From a mathematical point of view, the formal principle of MI might be seen as a precise formulation of quasi-induction. From a cognitive point of view, on the other hand, the gap between the two is considerable: while the latter deals with a local *inference step*, the former deals with a global *inference form*, as I will now explain.

Inference Step versus Inference Form

In quasi-induction the conviction that $P(n)$ is true for any given natural number n stems from one's ability to imagine starting from $P(1)$ and going through the inference steps, $P(1) \Rightarrow P(2)$, $P(2) \Rightarrow P(3)$, . . . ,$P(n - 1) \Rightarrow P(n)$. This does not mean that one actually runs through many steps, but that he or she realizes that in principle this can be done for any given natural number n. In particular, in quasi-induction one views the inference, $P(n - 1) \Rightarrow P(n)$, just as one of the *inference steps*—the last step—in a sequence of inferences that leads to $P(n)$. In MI, on the other hand, one views the inference, $P(n) \Rightarrow P(n + 1)$, as a variable *inference form*, a placeholder for the entire sequence of inferences. Consequently, he or she has no need to run through the specific inference steps, $P(1) \Rightarrow P(2)$, $P(2) \Rightarrow P(3)$, . . . ,$P(n - 1) \Rightarrow P(n)$, for the instantiation of these steps into the inference form, $P(n) \Rightarrow P(n + 1)$, are viewed as a logical necessity. Thus, while quasi-induction and MI are both instantiations of the transformational proof scheme, the latter is an abstraction of the former.

The Introduction of the Formal Principle MI in the NT Experiment

Unfortunately, at the time we conducted the NT experiment, we were unaware of the cognitive gap between quasi-induction and the formal principle of MI. We viewed the latter merely as a concise formulation of the former, and so was it presented to the students.

The students were accustomed to gradual reformulation of mathematical ideas—a practice we established in each of our teaching experiments. With a few exceptions, our students' ability to express mathematical ideas properly was initially very limited. Specific interventions were needed to bring about improvement in the quality of their mathematical exposition. The following are two of the interventions.

Individual students are asked to see the instructor to explain or articulate some of their homework or exams. In these individual meetings the student is helped to recognize that although her or his idea is correct or partially correct, it is expressed deficiently or insufficiently. The recognition often occurs when the instructor presents an interpretation of the student's answer that the student deems legitimate, yet initially unintended by her or him. Once the student arrives at the recognition of the need to modify her or his work, she or he is offered the option to rewrite the respective assignment, and upon completion the initial grade is adjusted according to the quality of the change.

The other intervention is that each instructional unit in the teaching experiment ends with an assignment in which students are asked to summarize what they perceive as the most important ideas of the unit. With these summaries in mind, and with further input from the students, the instructor then gives a presentation in which the main ideas learned are expressed in standard mathematical language. Following this, the students are assigned a set of review problems aimed at helping them solidify the ideas learned in terms of the final formulations.

It was in the context of the latter classroom practice that the formal principle of MI was presented as a reformulation of quasi-induction. The formal statement presented was the following:

$P(n)$ is true for all positive integers n if

1. $P(1)$ is true
2. For any positive integers k, whenever $P(k)$ is assumed true, $P(k + 1)$ is also true.

This was the first time in the course that the term *mathematical induction* was used; none of the above terms, including the term *quasi-induction*, had ever been used in class. Following this, the students were assigned a set of review problems consisting of combinations of implicit and explicit recursion problems in a variety of contexts: algebra, geometry, and calculus. The set also included nonrecursion MI problems (to be discussed in the next subsection).

Unlike the previous two stages, no significant classroom discussion ensued from this presentation. The students seemed to have easily assimilated the principle of MI into their scheme of quasi-induction: About 75% of the problems assigned were solved correctly and in terms of the principle of MI; the rest were either solved by quasi induction or by other means (see the next subsection below). The following is an example of one student's solution by MI to the problem, "Prove that a set of n elements has $2^n - 1$ nonempty subsets."

> Consider the case for $n = 2$, set $= \{a_1, a_2\}$
> Nonempty subsets $= \{a_1\}, \{a_2\}, \{a_1, a_2\}$
> Now we can form the subsets for $n = 3$ by adding
> (i) the subset of $n = 2$, $= \{a_1\}, \{a_2\}, \{a_1, a_2\}$
> (ii) the subsets of $n = 3$, where each now includes element
> a_3: $\{a_1, a_3\}, \{a_2, a_3\}, \{a_1, a_2, a_3\}$ and the subset $\{a_3\}$
> Examining this process, we see that if a set of n elements has S_n subsets, a set of $n + 1$ elements has $2S_n + 1$ has subsets.
> Now assume $S_n = 2^n - 1$
> Then $S_{n+1} = 2S_n + 1 = 2(2^n - 1) + 1 = 2^{n+1} - 2 + 1$
> $= 2^{n+1} - 1$
> Also, for $n = 1$, $S_1 = 1 = 2^1 - 1$
> By induction, $S_n = 2^n - 1$ for all n.

The nearly effortless reconceptualization of quasi-induction into MI by the students is attributed to the conceptual foundations cemented in Phases 1 and 2. The *kind* and *order* of the problems the students dealt with in these phases constituted two critical intellectual needs for them: The first is the intellectual need to *compute* a solution to problems that were intrinsic, not alien, to them. By "compute," it is meant developing a process pattern generalization to explain (in Problem 1) the reason Z is an upper bound for the sequence. The second is the intellectual need to *formulate* this process—making it explicit to themselves—in order to persuade others of the validity of their solutions. Quasi-induction, therefore, marks a vital developmental step toward the concept of MI.

Essentiality of MI

Beyond the level of abstracting the formal principle of MI, there is yet another level the NT teaching experiment did not set as a goal. It is the understanding that from a formal, axiomatic perspective, any proof of an assertion on the set of positive integers must use MI. This level of understanding can be reached only when the students possess the *axiomatic* proof scheme—the understanding that a mathematical justification must, in principle, have started originally from undefined terms and axioms. Although the axiomatic proof scheme should be an explicit goal of undergraduate mathematics curricula, it was a unrealistic goal in the NT teaching experiment, given the state of knowledge and mathematical maturity of the participants.

That the NT students did not reach this level of understanding can be seen in their responses to problems such as:

10. Prove that for any positive integer n, $\log (a_1 \cdot a_2 \cdot \ldots \cdot a_n) = \log a_1 + \log a_2 + \ldots + \log a_n$
11. Prove that for any positive integer $n \geq 4$, $2^n \leq n!$
12. Prove that the sum of the measure of the interior angles in a convex n-gon is $180° (n - 2)$.

Such problems introduced to many students no need to search for consecutive relations, because, they argued, "the problems can be solved in an easier way." Most of these students offered solutions to these problems that were not based on MI. For example, Problem 10 appeared in succession to Problem 9. Most of the students solved Problem 9 by MI but argued that Problem 10 can be solved in an easier way, as did one of the students:

> Let $y = \log a$
> We can say $a = 10^y$ by definition of log
> Now we can express $a_1 \cdot a_2 \cdot \ldots \cdot a_n$ as $10^{y_1} \cdot 10^{y_2} \cdot \ldots \cdot 10^{y_n} = 10^{y_1 + y_2 + \ldots + y_n}$.
> Therefore,
> $\log(a_1 \cdot a_2 \cdot \ldots \cdot a_n) = \log(10^{y_1 + y_2 + \ldots + y_n}) = (y_1 + y_2 + \ldots + y_n) \log 10 =$
> $= (y_1 + y_2 + \ldots + y_n)1 = y_1 + y_2 + \ldots + y_n$
> $= \log a_1 + \log a_2 + \ldots + \log a_n$

Problem 11 in particular was viewed as trivial, as one that needs no MI. For example, one of the students responded to this problem by presenting 2^n and $n!$ on the top of each other as follows:

$$2^n = \underbrace{\underbrace{2\cdot2\cdot2\cdot2}_{16}\cdot2\cdot\ldots\cdot2}_{n \text{ times}}$$

$$n! = \underbrace{1\cdot2\cdot3\cdot4}_{24}\cdot5\cdot\ldots\cdot n$$

explaining that:

> The first factor in $n!$ [i.e., 24] is greater than the first [corresponding] factor in 2^n [i.e., 16], and the rest of the factors, 5, 6, 7, . . . , are [correspondingly] greater than the factors, 2, 2, 2, . . .

Figure 10.4.
A polygon representing an *n*-gon drawn by a student to explain why the sum of the measures of the interior angles in a convex *n*-gon is 180° (n-2).

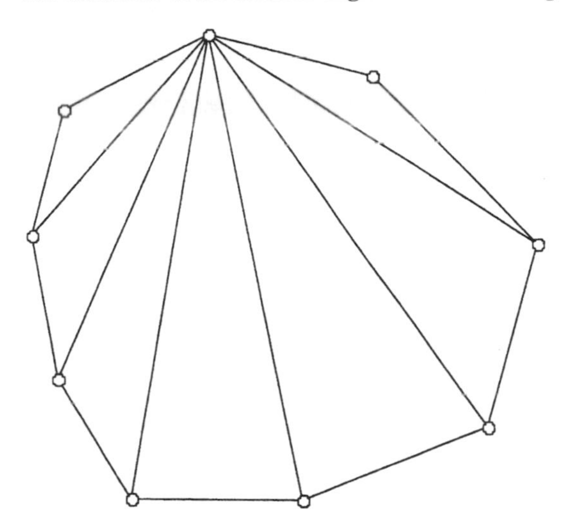

For Problem 12, one of the students presented his solution to the class by drawing a figure similar to Figure 10.4 on the board, explaining how the formula is derived from two facts: (a) in an n-gon there are n-2 triangles—pointing out that each vertex but two contributes one triangle—and (b) the sum of the measures of the angles in a triangle is 180°.

Following this student's presentation, the teacher presented a proof by MI of the same statement. The students protested that the MI proof is superfluous and that their classmate's proof makes more sense.

Although the classroom discussion on problems such as 10–12 helped students better see the underlying structure of all the problems they had dealt with in the three phases, it did not enhance their understanding beyond the level of MI as a transformational proof scheme. It did confirm, however, students' view of the nonessentiality of MI in solving certain kinds of problems.

CONCLUSION: THE DNR SYSTEM AS A CONCEPTUAL FOUNDATION FOR THE ALTERNATIVE INSTRUCTIONAL TREATMENT TO MI

MI is a significant proof technique in all fields of mathematics. Beyond its mathematical significance it provides a context to enhance students' proof schemes. Hence, I believe it must be part of secondary school mathematics curricula, especially those taught to college-intended students.

This chapter addresses two questions. The first two will be summarized, in turn, in the next two subsections. To these I will add a third question about the conceptual foundation of the alternative instructional treatment to MI. The latter will be addressed in the third, and final, subsection.

Question 1: How Are Students' Conceptions of and Difficulties with MI in a Standard Instructional Treatment a Manifestation of Their Proof Schemes?

The three most prevalent proof schemes among students are the authoritative proof scheme, the symbolic, nonquantitative proof scheme, or the empirical proof scheme. These ways of thinking impact and even determine students' understanding of MI, which, as we have seen, is conceived mainly as a result pattern generalization. Process pattern generalization—an expression of the transformational proof scheme—is virtually absent among students.

One possible cause of this situation is the standard instructional treatment of MI. In particular, the introduction of the principle of MI in this treatment is abrupt. Students do not see how the principle is born out of a need, an intellectual need that is, to solve problems that are intrinsic[12] to them. Consequently, they use the principle of MI as a prescription; thus reinforcing the authoritative and symbolic, nonquantitative proof schemes.

MI problems in standard textbooks are sequenced according to what their authors seem to perceive as an increasing order of difficulty, not in accordance to students' conceptual development. As a consequence, the task to use MI in all problems discriminately—including those for whom the students have alternative solutions—is alien[13] to these students and exacerbates their poor view of mathematics.

Question 2: What Are Students' Difficulties with and Conceptual Development of MI in an Alternative Instructional Treatment?

The new instructional treatment of MI takes into consideration the possible causes for the failure of the standard approach. In particular, it consists of phases corresponding to levels of conceptual development: It begins with the formation of quasi-induction, continues with its internalization and interiorization, and concludes with its abstraction into MI. Therfore, for the NT experiment students, the concept of quasi-induction seems vital in the process of learning the principle of MI. This observation is consistent with the historical development of MI, as it will be reported in Harel and colleagues (in preparation).

The most significant result reported in this chapter is that in this alternative treatment students changed their current ways of thinking, primarily from mere empirical reasoning—in the form of result pattern generalization—into transformational reasoning—in the form of process pattern generalization.

Question 3: What Is the Theoretical Basis for the Alternative Instructional Treatment Implemented in the NT Experiment?

The DNR system of pedagogical principles for designing, developing, and implementing mathematics curricula is the conceptual foundation for the alternative instructional treatment offered here. In the remainder of this chapter I present a synopsis of the system. A full description of the system appears in Harel (in preparation). The three essential pedagogical principles constituting the DNR system are the *duality principle, necessity principle*, and *repeated reasoning principle*.

The Duality Principle

Fundamental to the duality principle is the distinction between *ways of thinking* and *ways of understanding* (Harel, 1998, p. 497):

> Ways of thinking are students' apparatuses for filtering and interpreting what we intend to teach them. . . . They are distinguished from ways of understanding. A way of understanding is the meaning(s) students have for a specific concept. For example, students may understand the "derivative of a function" as the slope of a line tangent to the graph of a function, as the best linear approximation to a function near a point, etc., but they may understand it superficially (e.g., "derivative is x^{n-1} for x^n") or even incorrectly (e.g., "derivative is the quotient $\frac{f(x + h) - f(x)}{h}$").

The importance of this distinction is primarily in the mutual cognitive impact of ways of thinking and ways of understanding, which is the content of the duality principle (Harel, 1998): "*Students' ways of thinking impact their ways of understanding mathematical concepts. Conversely, how students come to understand mathematical content influences their ways of thinking.*"

By intervening in the students' ways of understanding specific content, in our case that of implicit recursion problems, the students altered their narrow ways of thinking about what constitutes proof—from authoritative, symbolic, non-quantitative, and empirical proof schemes into quasi-induction—a form of a transformational proof scheme. In turn, this new way of thinking impacted students' ways of understanding explicit recursion problems. This dual effect between ways of thinking and ways of understanding is an example of how the duality principle guided the design and implementation of the alternative instructional treatment of MI.

The Necessity Principle

The necessity principle—the second underlying conceptual basis for the alternative instructional treatments—asserts: "*Students are likely to learn if they see need for what we intend to teach them, where by 'need' it is meant intellectual need as opposed to social or economic need*" (Harel, 1985, 1990, 1998).

The term *intellectual need* refers to a behavior that manifests itself internally with learners when they encounter a situation that is incompatible with, or presents a problem that is unsolvable by, their existing knowledge. Such an encounter is intrinsic to the learners, for it stimulates a desire with them to search for a resolution or a solution, whereby they might construct new knowledge. There is no guarantee that the learners construct the knowledge sought or any knowledge at all, but whatever knowledge they construct is meaningful to them in that it is integrated within their existing cognitive schemes, because it is a product of an effort that stems from and is driven by *their personal intellectual need.*

The implementation of the necessity principle involves (a) recognizing what constitutes an intellectual need for a particular population of students relative to the concept to be learned; (b) developing a system of problem situations that correspond to their intellectual need, and from whose solutions the concept can be elicited; and (c) creating an instructional environment in which the student can elicit the concept through engagement with the system (Harel, 1998). These are not three steps of a recipe to be carried out chronologically; rather, these constitute, respectively, the essence of three inseparable aspects of research in learning, curriculum development, and teaching.

In Harel (1998), I distinguished among three kinds of intellectual needs in the learning of mathematics: the *need for computation*, the *need for formulation*, and the *need for elegance*. Relevant to the work here are the first two needs.

The need for computation means finding unknown objects or determining their causes. Examples of this need include (a) finding the solution of a system of equations, the GCD of two integers, the probability of an event, the cardinality of a set; and (b) determining the cause[14] (i.e., reason) for a system $Ax = b$ to have a solution or a unique solution.

The need for formulation is the need to communicate ideas, including convincing others of the truth of an observation. In the NT teaching experiment, a didactical contract (in the sense of Brousseau, 1997) was established according to which persuasion, not only ascertainment, is an essential part of proving. This practice induced students to reflect on and formulate quasi-induction—a significant step in the development of students' conception of MI.

The need for computation and the need for formulation were combined in the NT teaching experiment as a means to construct the concept of MI. For example, the task to find an upper bound fits well into what the NT students conceive as doing mathematics. From this reason the need to compute an unknown quantity, or determine conditions for its existence, is well understood and appreciated by them. Once this need was established, the need for formulation—that is, the need to persuade others—facilitated the institutionalization of the superiority of the process pattern generalization over the result pattern generalization.

Figure 10.5.
The DNR system.

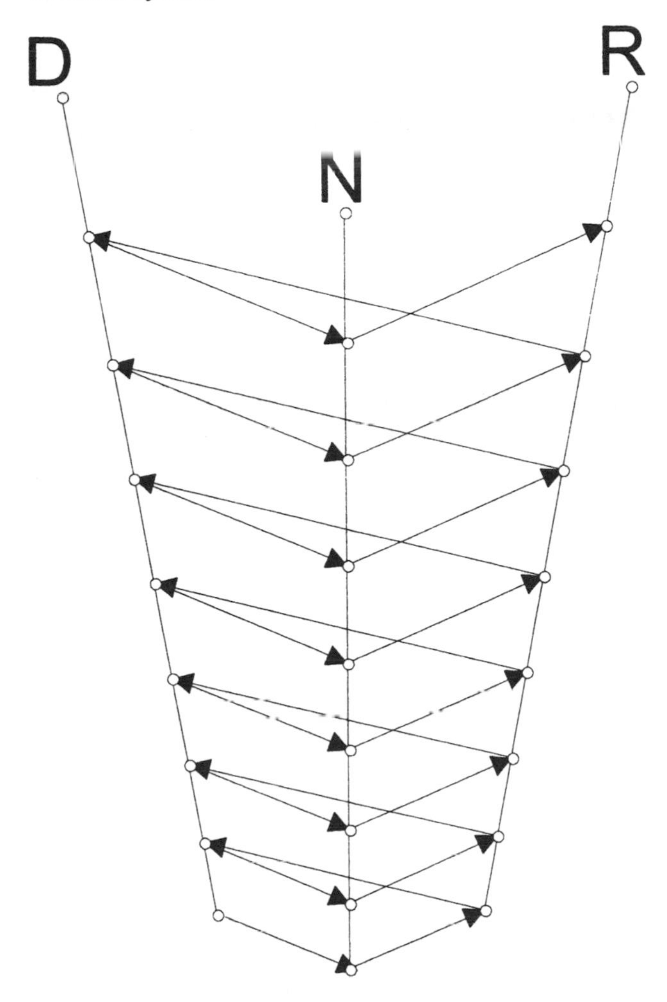

The Repeated Reasoning Principle

While the necessity principle aims at the initial formation of a concept, the repeated reasoning principle addresses its internalization and interiorization. The principle asserts: *Students must practice reasoning in order to internalize and interiorize specific ways of thinking and ways of understanding*. The *repeated* attempt by the NT experiment students to attend to the underlying common structure to the implicit recursion problems and the explicit recursion problems helped them first internalize and then interiorize quasi-induction. In turn, the *repeated* reasoning in applying quasi-induction to solve these two types of problems help them assimilate the principle of MI almost effortlessly. Thus, repeated reasoning, not mere drill and practice of routine problems, is an important mechanism for solidifying concepts.

The DNR as a System

The three principles—duality, necessity, and repeated reasoning—constitute a system in the sense that they interdependently address three fundamental aspects of the learning–teaching process (Figure 10.5).

1. *Instructional objectives.* In designing, developing, and implementing mathematics curricula, ways of thinking and ways of understanding have been our ultimate cognitive objectives. We have addressed them simultaneously, for one affects the other.
2. *Formation of concepts.* Meaningful concepts can only be elicited through solutions to problems that correspond to students' intellectual need (the necessity principle). The concepts elicited as well as the ways in which they are elicited constitute and at the same time impact students' ways of thinking and ways of understanding.
3. *Internalization/interiorization of concepts.* Ways of thinking and ways of understanding are internalized and interiorized through repeated solutions to fundamental problems.

NOTES

1. This chapter focuses on cognitive and didactical considerations of mathematical induction in two instructional treatments: one standard, the other new. Initially no historical-epistemological factors were considered in designing and implementing the alternative treatment. However, recent historical analysis of MI has revealed a pleasantly surprising observation: the path of conceptual development in the alternative treatment parallels the historical development of MI. This observation will be reported in a sequel to this chapter (Harel, Manaster, McClure, & Sowder, in preparation).

2. I thank Alfred Manaster, Maria Alessandra Mariotti, James McClure, and Larry Sowder for the many stimulating conversations on issues addressed in this chapter.

3. The research presented here is part of the PUPA (Proof Understanding, Production, and Appreciation) Project, supported in part by a grant from the National Science Foundation to Larry Sowder of San Diego State University and the author of this chapter. Opinions expressed are those of the authors and not necessarily those of the Foundation.

4. See note 3.

5. The term "scheme" is used here in the sense of Piaget: "The part that is common to applications or repetitions of the same action" (Vuyk, 1981, p. 63). Glasersfeld (1991, pp. 55–56) specifies the functionality of schemes: "Schemes are composed of three elements: (1) an initial experiential item or configuration (functionally linked to what the observer would categorize as 'trigger' or 'stimulus'), (2) an activity the subject has associated with it, and (3) a subsequent experience associated with the activity as its outcome or result." The essential actions in a proof scheme are "persuasion" and "ascertainment." These actions are applied in situations one desires to evaluate hypotheses. The kind of persuasion and ascertainment one associates with such situations depends

on her or his experience of the outcome actions applied. For example, the inductive proof scheme is most likely the result of the everyday experience in which people's evaluation of hypotheses is probabilistic in nature.

6. For a discussion on the ineffectiveness of traditional analogy-based instruction, including the domino analogy to MI, see Greer and Harel (1998).

7. Any algebraic expression containing an ellipsis is a recursive representation of a function.

8. Some readers of Harel and Sowder (1998) thought that transformational proof scheme refers only to students' arguments using visual, spatial motions. There is no condition in the definition of transformational proof scheme that restricts its applicability to geometric situations. In fact, both in Harel and Sowder and here, transformational proof scheme is demonstrated with algebraic examples. In the former publication, however, most of the examples dealt with geometry and graphs, which might be the cause of the misinterpretation.

9. Editors' Note: The author uses "interiorization" here in a different sense than it is used in APOS theory.

10. Fibonacci numbers, $f_0, f_1, f_2, \ldots$ are defined by the equations $f_0 = 0, f_1 = 1$, and $f_n = f_{n-1} + f_{n-2}$.

11. In presenting the Tower of Hanoi problem, the instructor suggested to the students to first think in terms of the specific number of disks.

12. Generally speaking, an "intrinsic problem" is one the student understands and appreciates. A specific characterization of this term is in Harel (in preparation).

13. Generally speaking, an "alien problem" is one that is not intrinsic. A specific characterization of this term is in Harel (in preparation).

14. For more discussion on the notion of *causality*, see Harel (1999).

REFERENCES

Baxandall, P. R., Brown, W. S., Rose, G. St. C., & Watson, F. R. (1978). *Proof in mathematics*. Keele, UK: Institute of Education, University of Keele.

Boyer, C. (1968). *A history of mathematics*. New York: Wiley.

Brousseau, G. (1997). *Theory of didactical situations in mathematics*. Dordrecht: Kluwer Academic.

Cobb, P., & Steffe, L. (1983). The constructivist researcher as teacher and model builder. *Journal for Research in Mathematics Education, 14*, 81–94.

Dubinsky, E. (1986). Teaching of mathematical induction I. *Journal of Mathematical Behavior, 5*(3), 305–317.

Dubinsky, E. (1989). Teaching of mathematical induction II. *Journal of Mathematical Behavior, 8*, 285–304.

Ernest, P. (1984). Mathematical induction: A pedagogical discussion. *Educational Studies in Mathematics, 15*, 173–189.

Glasersfeld, E. v. (1991). Abstraction, re-presentation, and reflection: An interpretation of experience and Piaget's approach. In L. P. Steffe (Ed.), *Epistemological foundations of mathematical experience* (pp. 45–67). New York: Springer-Verlag.

Greer, B., & Harel, G. (1998). The role of analogy in the learning of mathematics. *Journal of Mathematical Behavior, 17*, 5–24.

Harel, G. (1985). *Teaching linear algebra in high school*. Unpublished doctoral dissertation, Ben-Gurion University, Israel.

Harel, G. (1990). Using geometric models and vector arithmetic to teach high-school students basic notions in linear algebra. *International Journal for Mathematics Education in Science and Technology, 21*, 387–392.

Harel, G. (1998). Two dual assertions: The first on learning and the second on teaching (or vice versa). *American Mathematical Monthly, 105*, 497–507.

Harel, G. (1999). Students' understanding of proofs: A historical analysis and implications for the teaching of geometry and linear algebra. *Linear Algebra and Its Applications*, 302–303, 601–613.

Harel, G. (in press). Students' proof schemes revisited: Historical and epistemological considerations. In P. Boero (Ed.), *Theorems in shcool.* Dordrecht: Kluwer.

Harel, G. (in preparation). PUPA's two complementary products: Conceptual framework for Proof Schemes and DNR system for enhancing students' proof schemes.

Harel, G., Manaster, A., McClure, J., & Sowder, L. (in preparation). *Mathematical induction as a proof scheme: Parallels between students' conceptual development and historical epistemology.*

Harel, G., & Sowder, L. (1998). Students' proof schemes. *Research on Collegiate Mathematics Education*, Vol. III, 234–283.

Show, B (1978). *n* and *S. Mathematics Teaching, 82*, 6–7.

Tall, D. (1992). The transition to advanced mathematical thinking: Functions, limits, infinity, and proof. In D. S. Grouws (Ed.), *Handbook of research on mathematics teaching and learning* (p. 497). New York: Macmillan.

Vergnaud, G. (1988). Multiplicative structures. In J. Hiebert & M. Behr (Eds.), *Number concepts and operations in the middle grades* (pp. 141–161). Reston, VA: National Council of Teachers of Mathematics.

Vinner, S. (1985). Concept definition, concept image, and the notion of function. *International Journal of Mathematics Education, Science and Technology, 14*, 293–305.

Vuyk, R. (1981). *Overview and critique of Piaget's genetic epistemology 1965–1980.* London: Academic Press.

Woodall, D. (1981). Finite sums, matrices and induction. *Mathematical Gazette, 65*, 92–103.

11

Reflections on Mathematics Education Research Questions in Elementary Number Theory

Annie Selden and John Selden

This monograph intentionally raises more questions than it answers. Indeed the editors' aims included convincing readers that many interesting questions in the teaching and learning of number theory await their attention, and that there are engaging number theory tasks with which to investigate students' mathematical sense making.

Number theory appears to be a rather neglected area in the mathematics education research literature. Whether one looks at the *NCTM Handbook of Research in Mathematics Teaching and Learning* (Grouws, 1992) or the NCTM Yearbook, *Developing Mathematical Reasoning in Grades K–12* (Stiff, 1999), one finds almost nothing on students' engagement with number theory topics. By number theory we mean, as do the monograph editors, results concerning the structure of the integers that are not primarily computational, for example, questions about factorization or the distribution of primes. Whatever the reasons for this research lacuna, this monograph, devoted entirely to various number theory investigations and reports, should help fill this void and heighten the community's awareness of the potential for fruitful investigations. These might concern individual students' cognitions on a variety of topics from divisibility and prime factorization to number theoretic generic proofs and mathematical induction, or they might be more sociocultural studies.

The authors of this monograph approach topics concerning the teaching/learning of number theory from a broad range of perspectives and investigate university students having various levels of mathematical sophistication from preservice elementary teachers to second-year Cambridge University mathematics majors. Running across the contributions, however, are several common threads upon which we comment: the potential of number theory for the teaching and learning of problem solving, reasoning, and proof; questions regarding the language and images of divisibility; philosophical stances taken; theoretical frameworks used; and implications for teaching. In the spirit of the monograph and drawing on it, we focus primarily on questions and new directions for investigation, some ranging well beyond number theory itself.

PROBLEM SOLVING, REASONING, AND PROOF

Elementary number theory seems an ideal place to learn about problem solving, reasoning, generalization, abstraction, and proof because students from middle school through college feel comfortable dealing with whole numbers. They are much more familiar with the positive integers than with rational, real, or complex numbers. Appropriately chosen elementary number theory problems can provide an excellent opportunity for developing and evaluating mathematical arguments. The NCTM *Principles and Standards* (2000) calls for such problem solving and argumentation in both the Reasoning and Proof and the Number and Operations Standards for Grades 6–8. Middle school students could, for example, explain why the sum of the digits of any multiple of 3 is itself divisible by 3 (National Council of Teachers of Mathematics, 2000, p. 217). Also the Number and Operations Standard for Grades 9–12 explicitly states that all students should "use number-theory arguments to justify relationships involving the whole numbers." One could ask: What can you say about the number that results when you subtract 1 from the square of an odd integer? Students could easily verify that the number is divisible by 4 and conjecture that it is divisible by 8. Proving this would involve representing the number as $(2n + 1)^2 - 1 = 4(n)(n + 1)$ and recognizing that one of n or $n + 1$ is even (NCTM, 2000, pp. 290–292).

Teppo's vignette (chapter 6, this volume) shows how students can begin reasoning about the number of divisors of small positive integers. Then, with scaffolding by the teacher along with a suitable didactic object (a divisor table), students can conjecture rules and explain why they hold. This can lead to subsequently considering the general question of how to determine the number of divisors of any positive integer, given its prime factorization. In general, however, care should be taken by teachers to ascertain whether a proof is obtainable by elementary means. It is a well-known "folk theorem" among mathematicians that number theory questions and conjectures can be easy to state but phenomenally hard to prove (e.g., about partitions or the distribution of primes). While the ordered partition problem (Rowland, Chapter 9, this volume) is a potentially rich classroom task, finding regularity in numbers of (unordered) partitions is subtle and difficult and involves an understanding of modular-forms. (See Peterson (2000) for a description of mathematician Ken Ono's recent results on partition congruences.) A collection of elementary number theory problems whose solutions/proofs are accessible by elementary means, along with advice for teachers on potential student "pitfalls," would be a useful pedagogical aid.

Why might elementary number theory be a good place for university students to learn about proofs? Apparently less abstraction is required to construct elementary number theory proofs than, say, abstract algebra proofs, and this can be an advantage for students with a weak conception of proof. Currently in the United States, university mathematics majors, including preservice secondary teachers, can often avoid dealing with proofs until a "bridge" course on proofs, often followed by an abstract algebra course. Such students' entire previous experience may lead them to believe that mathematics deals mainly with numbers and functions. Suddenly, in an abstract algebra course they need to treat operations abstractly and see commonalties among radically different examples—a cause of considerable confusion.

Recently, while teaching abstract algebra to preservice secondary teachers and mathematics majors, we observed some difficulties with abstraction. The students were presented with various sets, together with operations. They were asked to determine whether these were semigroups and contained an identity, a zero, or other idempotents. The sets were subsets of the reals, 2×2 real matrices, or real functions. The students "knew" that juxtaposition stood for the (appropriate) operation, yet several students worked with juxtaposition as if it were ordinary multiplication. For example, the condition for a zero element a of a semigroup S (for all x in S, $ax = a$ and $xa = a$) was not readily rewritten in terms of other operations (e.g., as $a + x = a$ and $x + a = a$ for addition, or as $(f \circ a)(x) = a(x)$ and $(a \circ f)(x) = a(x)$ for function composition). One student said, "When I see this [juxtaposition], I think it's multiplication." For these students, juxtaposition should have been, but was not, functioning as a visual metaphor, that is, suggesting the operation worked a little, but not exactly, like ordinary multiplication. It has been observed that in trying to cope with abstract algebra questions, university students often attempt to reduce the level of abstraction by "folding back" to what they know about numbers, thereby changing an unfamiliar problem into a more familiar one (Hazzan, 1999). This kind of difficulty with abstraction often occurs while students are also struggling to construct a conception of proof that is consistent with the standard (mathematicians') concept. We suspect these two difficulties (concerning abstractions and the nature of proof) interact, exacerbating both.

By contrast, in elementary number theory, students deal with objects (integers) and operations (ordinary multiplication and addition) that are familiar to them. Hence, they can concentrate on discovering and constructing proofs without being distracted by simultaneously having to extend their conceptions of the operations and objects they are studying. An example of an elementary number theory statement not involving excessive abstraction is the following: A sequence of positive integers p, $p + 2$, $p + 4$ is called a *triple of primes* if and only if all three of p, $p + 2$, $p + 4$ are primes. The only triple of primes is 3,

5, 7. Our "bridge" course students have found this statement challenging, but possible to prove. One student said, "The triple of primes problem kept me up until two in the morning. I was in bed staring at the ceiling when inspiration struck around four in the morning. It's not the kind of thing you do in calc."

Rowland (this volume) makes a convincing case for the use of generic proofs in number theory but also notes (note 12) that sometimes generic presentations seem ruled out. One question to ask from a pedagogical point of view is whether a specific generic proof is likely to be illuminating, even if one can easily find a suitable particular case. For an example at the school level, consider the proof that subtracting 1 from the square of an odd integer yields a number divisible by 8. It is hard to see that a generic proof would be better understood beginning perhaps with 11 (instead of $2n + 1$) and writing $11^2 - 1 = (2 \cdot 5 + 1)^2 - 1 = 4 \cdot 5^2 + 4 \cdot 5 + 1 - 1 = 4 \cdot (5^2 + 5) = 4 \cdot 5 \cdot (5 + 1)$. Yet this seems to follow Rowland's first four principles. The particular case, 11, is neither trivial nor complicated. After one has written $11 = 2 \cdot 5 + 1$, one can track 5 through the stages of the argument. Invariant aspects of the proof transfer to values of n other than 5, and the reasoning is constructive. By contrast, a generic proof of the digit-sum rule for divisibility by 9 obtained by writing $864 = 8 \cdot 100 + 6 \cdot 10 + 4 = (8 \cdot 99 + 6 \cdot 9) + (8 + 6 + 4) = 9(8 \cdot 11 + 6) + (8 + 6 + 4)$ seems both illuminating and generalizable. At a more advanced level, Rowland has developed an enlightening generic version of an elementary and elegant proof that the Euler ø-function is multiplicative. On the other hand, we rewrote Bolker's (1970) textbook proof of the same result using $m = 3$ and $n = 8$ and concluded it would not be illuminating. Thus not every proof seems suitable, *ex post facto*, for treatment via generic example, even following Rowland's principles, and we join him in calling for a refinement of those principles.

On an individual level, the effectiveness of generic proofs depends on a student's ability to see the features of the general argument in the particular case, that is, on a student's ability to make structural generalizations (Rowland, referring to Bills [1996]). Structural generalization is a way of thinking similar to Harel's (Chapter 10, this volume) process pattern generalization for mathematical induction problems, in which students look for a pattern relating a statement about one integer to the (same) statement about the next (or previous) integer. Just one of Harel's student groups seems to have, on their own, come upon process pattern reasoning for the sequence $\sqrt{2}, \sqrt{2+\sqrt{2}}\ \sqrt{2+\sqrt{2+\sqrt{2}}} \ldots$; a way of thinking subsequently encouraged by the instructor. Not all students seem to make such generalizations naturally. Are there ways to help students become better at it? Harel suggests the importance of repeated reasoning, the R in his DNR system. A carefully structured sequence of problems together with suitable interventions by the teacher to encourage internalization and interiorization (à la Harel) might well be helpful. Further studies could help provide details that might help teachers engender such reasoning in their students.

In summary, why is number theory ideal for introducing students to reasoning and proof? (1) Students get to deal with familiar objects, thereby reducing the level of new abstraction and the concomitant disequilbration; (2) when suitably selected, such proofs are accessible, that is, students need only reason from first principles together with a certain amount of ingenuity; and (3) often number theory proofs have generic versions, allowing students to see, or even to prove on their own, a general result after having considered a suitable particular case.

QUESTIONS REGARDING THE LANGUAGE AND IMAGES OF DIVISIBILITY

Within mathematics, the concept of divisibility in the integers seems somewhat unusual in having such a wide assortment of terms to describe different aspects of essentially the same mathematical phenomenon: A is *divisible* by B, B *divides* A, B is a *factor* of A, B is a *divisor* of A, and A is a *multiple* of B. In some textbooks for preservice teachers (e.g., Billstein, Libeskind, & Lott, 1984) and for discrete mathematics (e.g., Epp, 1990), it is simply stated that these five phrases are alternative ways of expressing the fact that there is a unique integer C such that A = BC. But why are there so many different terms? What aspects of the concept does each highlight? To what extent does the mere existence of so many essentially equivalent terms cause students confusion? How do students cope with this variety of terms? What is the effect of students' own alternative terms, for example, B *divides* A *evenly*? We speculate on some of these issues, but definitive answers need to come from research.

The words *divisible, divides*, and *divisor* emphasize those integer division situations in which the remainder, R, in the division algorithm, $A = BC + R$, $0 \leq R < B$, is zero. In the case of *divisible*, A's role is highlighted, whereas in the case of *divides* and *divisor*, B's role is stressed. The terms *factor* and *multiple* emphasize two slightly different multiplication situations. In the case of *factor*, it is the factorization, and to some extent the prime factorization, of A that is being regarded, with B's role in that factorization stressed. However, with *multiple*, an Archimedean-like property (repeated addition) of the integers may be being stressed, with emphasis on the fact that $A = B + B + B + \ldots + B$ where there are C of the B's. Mathematically, the division algorithm, prime factorization, and the Archimedean property stress somewhat different features of the positive integers, but textbooks often only offer students a formal definition, that is, the simple requirement that there exist a unique C such that A=BC, together with illustrations and exercises. Yet in selecting one term over another, it may sometimes appear that division is being stressed, while at other times, multiplication. Also, sometimes A's role is emphasized, other times B's. This, in itself, might cause students some confusion. This is not to say that beginning university students should be introduced to all these nuances, but only that there's probably more here than just five ways of saying exactly the same thing and that the terminology used may influence what comes to mind at various times during cognition.

In attempting clarification of tasks such as "Is $M = 3^3 \times 5^2 \times 7$ divisible by 15?" or "What is the remainder if you divide $6 \times 147 + 1$ by 6?", students seem to be trying to make sense of the question in their own everyday language by asking the interviewers whether "divides evenly" or "what's left over" is wanted (Campbell, Chapter 2; Brown, Thomas, & Tolias, Chapter 3; Zazkis, Chapter 4, this volume). This suggests that some students may be bringing to mind various visual images of division or multiplication. These may be partitive or quotitive as suggested by Campbell (this volume), but might also be of other kinds. For example, might the terms *quotient* and *remainder* bring to mind visual images such as

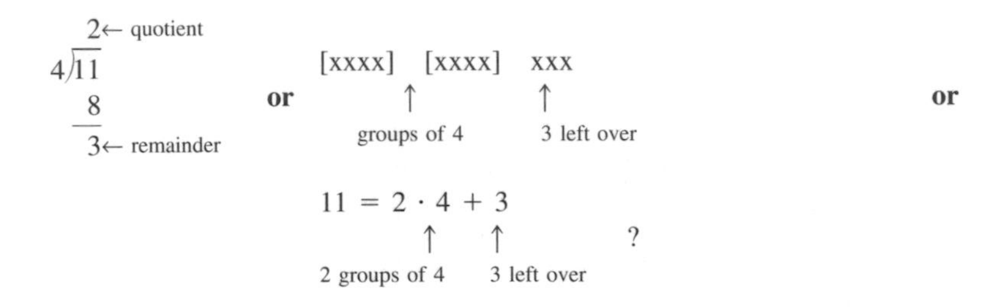

In the first of these, it is possible to interpret the quotient as what sits on top and the remainder as what's left at the bottom. One discrete mathematics textbook author uses these three illustrations to introduce the division algorithm, apparently calling on students' everyday conceptions of "left over" (Epp, 1990, p. 173). It would be an interesting research question to see what sorts of mental images of the division algorithm, or of prime factorization, that university students bring with them and to what the degree they are aware of using inner vision or inner speech.

When one asks students such questions as "Is $M = 3^3 \times 5^2 \times 7$ divisible by 15?" or "Is $3^4 \times 5^5 \times 7^3 \times 13^{18}$ a multiple of M?", not only is one asking them to discriminate between multiplication and division and between integer and rational division, but one is also introducing additional representational complexities. While the question "Is $M = 3^3 \times 5^2 \times 7$ divisible by 5?" seemed relatively easy for many students like Alice because they could see the 5 "in there" (Brown et al., this volume), the question of divisibility by 15 often found them resorting to their calculators to find M and divide by 15 to see whether "it" would "go in evenly." Would asking whether $M = 3^2 \times 5 \times 7 \times 15$ is divisible by 15 have elicited the same tendency to calculate M? We think not. This resort to an action (procedure) seems to be an instance of "folding back" to the Primitive Knowing level (Pirie & Kieren, 1994; Towers, Martin, & Pirie, 2000), that is, to a portion of one's mathematical knowledge with which one is comfortable and secure—a way to deal with the complexity of the situation. In addition, for preservice elementary teachers, it may be that their view of doing mathematics consists largely of performing computations to get answers, together with a propensity to view symbols such as + and × as commands to calculate, further influencing their tendency toward a computational interpretation of such tasks. Furthermore, to determine whether a specific number not given in prime factored form, such as 344578, is divisible by another number, say 457, it seems quite natural to resort to long division or a calculator. While the formal definition of "B divides A" is especially good for proving results, such as M|N and N|P together imply M|P, it gives no means of deciding whether for specific integers there actually is a C such that A = BC.

We have noticed a slightly different image of divisibility used by some of our "bridge" course students when attempting to construct proofs in elementary number theory. They often want to convert "B divides A" into "$\frac{A}{B}$ is an integer." This leads to contemplating properties of pairs of numbers, for example, what can be said about a noninteger times an integer? Such reflections do not seem as straightforward as working with the equation $A = BC$, where A, B, C are integers. To see this, consider the following theorem: If M divides N and M does not divide P, then N does not divide P. The essence of one student's "proof" went as follows: "Since M divides N, there exists an integer K so that $N = MK$ and so $\frac{N}{M}$ is an integer. Since M does not divide P, $\frac{P}{M}$ is not an integer. Then $\frac{P}{N} = \frac{P}{MK}$ is not an integer. So N does not divide P." To complete the proof, the student would have needed to argue that a noninteger such as $\frac{P}{M}$ divided by an integer such as K is a noninteger. This image of divisibility results in the consideration of an object, $\frac{A}{B}$, but one not as easily handled as the object $A = BC$, for which most students know transformation rules. It would be interesting to investigate implications of this alternative image of divisibility.

Sometimes in an attempt to answer questions conceptually, rather than procedurally by resorting to their calculators, students in the interviews seemed to be using their concept images, some that included an idea of "taking something out" (Campbell, this volume). There seem to be several possible interpretations of this idea. One could "take 3×5 out" of M to produce 15 as Riva and Lanny did and then have difficulty interpreting the role of 15 (Campbell, this volume). In the case of M, one "takes out" a factor of 15. However, there are other possible interpretations of "taking out." Considering the case of 11 divided by 4, one might also think of "taking 4 out" of 11 twice, using a repeated subtraction view of division, and having 3 "left over." Or, one might also "take 2 fours out" of 11 and have "3 left over." Depending which image of "take out" comes to mind, one might have a factor, a divisor, or a quotient, making it difficult to answer the question correctly. In what ways do phrases like "take out" and "what's left over" influence students' ways of looking at questions such as "Is $M = 3^3 \times 5^2 \times 7$ divisible by 15?" While students should learn to use mathematical terms precisely, they also need to unpack them and call on relevant mental images and generic examples. If a term like "left over" calls up an appropriate quotitive view of division, this would seem especially useful later for preservice teachers.

Many other issues regarding divisibility have been raised by various chapter authors, for example, the necessity to distinguish the indivisible units of integer division from the infinitely divisible units of rational division, as well as the importance of negotiating successfully between the modular ($2 \cdot 10 + 1$), fractional ($10\frac{1}{2}$), and decimal (10.5) representations of "21 divided by 2" (Campbell, this volume). All of these are ripe for further investigation.

PHILOSOPHICAL POSITIONS

In Chapter 1, under Epistemological Orientation, Campbell and Zazkis remark on the relevant philosophical views of the contributors to this monograph and, more generally, of the mathematics education research community. The material in the monograph, along with most research in mathematics education, reports on empirical observations that can largely be understood without referring to the underlying philosophical views of the authors. Nevertheless such views *do* play an important role in research because they influence the kinds of questions that will be, or even can be, asked as well as the kinds of studies regarded as providing acceptable warrants for answers to these questions.

Campbell and Zazkis correctly point out that the research community in mathematics education takes a constructivist view of learning. This entails two main ideas. First, individuals construct their own knowledge from their experiences and what they already know. Here, "construct" is not synonymous with "invent"—few people would claim students invent all of mathematics themselves. "Construct" also refers to the coordination and integration of new experiences with one's existing knowledge. Constructing knowledge is often thought to depend on some form of reflection, that is, thinking about one's actions (including mental actions or thoughts) and their effects. We note that reflection requires focusing one's attention and suspect one's focus (James, 1910) plays a large role in learning. Perhaps the importance of focus in learning can be most easily seen in situations involving its absence. Many people have had the experience of looking up a phone number, dialing it, and moments later not being able to remember it. No doubt their focus was on something other than the number, even though they were sufficiently conscious of it to dial it correctly. Like dialing a phone number, small mathematical tasks can be accomplished without much focus and it seems very unlikely that such situations could lead to the construction of lasting conceptual knowledge.

The role of focus in learning has not been well investigated in mathematics education, although it could be. For example, one might study it by examining the effects of tasks that incorporate distractions (i.e., that reduce focus) or by asking a number of students to report on what they are thinking about at various times. It would be interesting to know what students were focusing on during various kinds of lectures, group problem-solving sessions, and so on. It might also be interesting to know the effects of focus on independent study (homework).

The idea that knowledge is individually constructed leads to studies of how this occurs, that is, to studies of the mind. It provided legitimacy for research that had been largely blocked as late as the mid-twentieth century by behaviorism, the then dominant view in psychology. The idea was that science should be very reliable and predictive, rather like physics, and thus should depend only on directly observable phenomena. This led to concentrating on stimuli and responses, and eventually to the idea and study of conditioning. Such work was very reliable, but of limited use in explaining mathematical understanding, problem solving, learning, and so on. More importantly, the logical positivist views underlying behaviorism, which insisted on a level of reliability requiring direct observations, prevented serious examination of mental phenomena because such phenomena cannot be directly observed.

Returning to constructivism, the second major idea mentioned by Campbell and Zazkis (this volume) is that "there is no way of knowing about any metaphysical or mathematical realities independently of our own experience" or, more briefly, one cannot objectively know about things in the external world. Constructivism that includes this idea is called *radical constructivism* and only some of the mathematics education community ascribe to it. It is often difficult to see the effect of radical constructivism in individual research papers. Viewed as a whole, however, much of the literature has little to say about the relationship between an externally presented mathematical task and a person's moment-to-moment cognition in executing it. This apparent neglect of the boundary between "outside" and "inside" could be just a matter of chance, or it might derive from a feeling that a person cannot really "know" about the "outside" task or that all that really matters is the perception of the task—basically radical constructivist points of view. Seen in this way, radical constructivism may guide research to questions that do not consider the boundary between "outside" and "inside" and be limiting in a way similar to the way behaviorism was limiting some twenty or thirty years ago (although to a lesser degree). For example, there is essentially no work on the role of imitation in constructing procedural knowledge even though most of what is assessed in K–14 mathematics teaching is procedural and the tendency to imitate is a major feature distinguishing humans from other primates who favor emulation (Tomasello, 1997, 1999). Brown and colleagues' (this volume) interesting stage analysis of the construction of understanding from procedural knowledge illustrates this, in that nothing could be said about the origins of that procedural knowledge.

The central idea of radical constructivism, that is, that one cannot objectively know about the external world, follows from the fact that one has a single set of senses and, as a result, has nothing independent against which to test perceptions. Indeed, the perceptions themselves depend on one's own conceptions and knowledge, and thus inherently lack objectivity. While these points seem irrefutable, there is more to say. The important thing about human intelligence is not that people cannot objectively know about the external world (in an absolute sense), but rather that they can know about it (in an approximate way). Although people lack an automatic, biological way of objectively knowing about the external world, they do gradually develop mental mechanisms that can be used for reducing uncertainty about it. They are able to take several perspectives and can call on previous experience. Observations may be revisited and the results of their integration into a person's knowledge base can be questioned and adjusted (through Piagetian processes called disequilibration and reequilibration). For these mechanisms to yield a reliable, integrated, inner picture of (part of) the external world requires a way of drawing inferences that is about as reliable as deductive reasoning. There appears to be such a way, but it is not deductive reasoning itself because (two-valued) logic does not fit uncertainty very well.

Radical constructivism does not take account of this perspective and implicitly sees knowledge in an all or nothing way, that is, as expressed by statements that have only two truth values and that are combined using ordinary deductive reasoning and two-valued logic. This view of knowledge and reasoning is very widely accepted, especially in the West. However, in practice, people often appear to successfully use something like logic and deductive reasoning with statements that are not quite certain, that is, statements describing a kind of approximate knowledge that radical constructivism appears to neglect. We think approximate truth and reasoning can be precisely modeled. Doing so might suggest a more moderate view of radical constructivism and thereby enlarge the scope of research in mathematics education.

The uncertainty of statements can be viewed in two ways. One, which we will call *probabilistic uncertainty*, views statements as two-valued, even though the value may not be known. In the other, which we call *synthetic uncertainty*, the uncertainty is due to a synthetic definition or concept (as opposed to the precise, analytic kind of definition common in advanced mathematics). Synthetic definitions are at least partially incomplete descriptions of existing things and introduce a kind of inherent uncertainty into a statement. For example, if one says country M is "fairly democratic," one does not mean that out of ten copies of M, eight would be democratic and two would be dictatorships. One is instead (partly) expressing uncertainty of the exact meaning of "democratic." Indeed, in normal usage there might not *be* an *exact* meaning that is capable in every case of sharply distinguishing democratic from nondemocratic, that is, capable of supporting statements having only two truth values.

Often probabilistic uncertainty does not combine well with deductive reasoning because valid arguments can yield conclusions with probabilities much smaller than the probabilities of the premises. For example, in modus ponens, if the premises both have probability at least x, $0 \leq x \leq 1$, then one can only be sure the conclusion has probability at least $2x - 1$, for example, premises with probability at least 0.8 only yield a conclusion with probability at least 0.6. Thus, at this degree of probability, chains of several deductions yield conclusions with probabilities too small to be informative. However, often a person makes an observation of great (but not absolute) certainty. For example, a reader of this monograph may be quite certain that he or she is actually reading it, as opposed to dreaming, say with probability 0.99. This level of certainty works very well with modus ponens because premises with probability at least 0.99 yield a conclusion with probability at least 0.98. This is also true for modus tollens and a number of other valid arguments. The effect of this is that modest chains of deductions yield "almost certain" conclusions whenever the premises are "almost certain."

For synthetic uncertainty the situation is better. One might analyze this kind of uncertainty by assigning statements a numerical valve between 0 (false) and 1 (true), just as in the case of probabilistic uncertainty. But here it seems appropriate to take the value of the statement "p and q" to be the minimum of the individual values of p and q, the value of the statement "p or q" to be the maximum of the individual values, and the value of "not p" to be the one minus the value of p. These values together with the operations "and," "or," and "not" form a lattice with all of the properties of a Boolean algebra except the existence of complements, that is, "p or not p" might not have value 1, and "p and not p" might not have value 0. Using these values and operations, modus ponens and modus tollens and many other arguments valid in ordinary logic are valid in the sense that if the premises are approximately true (to *some* degree), then so is the conclusion (to the *same* degree). Indeed, any argument valid in the ordinary sense is also valid in this approximate sense if all of the statements used in it are assumed to be either approximately true or approximately false. In such situations, long chains of inferences are reliable, that is, do not lose truth value just as is the case with ordinary deductive reasoning.

The above description of approximate reasoning under small degrees of uncertainty suggests that a person can (tacitly) use it as part of the process of integrating new information into his or her knowledge base and can very reliably know about the external world, although perhaps not quite in the (two-valued, objective) sense referred to in radical constructivism. Thus there appears to be no philosophical impediment to investigations of the border between the external world and the inner mental world. In particular, the relationship between a mathematical problem written on paper and the corresponding inner speech version, that is normally produced when reading, can be investigated. To what degree is this correspondence "accurate"? For example, one of our college algebra students ignored the word "positive" in a test problem even after it was pointed out to him. To what degree, and why, do such discrepancies occur? What knowledge does a problem call to mind and by what mechanism? What part of solving a problem is conscious? What part is mental and what part is external?

When solving a linear equation a person may report nothing conscious between writing one line and writing the next. This suggests a significant part of the process is not conscious. In addition, if immediately after a line of the solution is written, that line is covered from the solver's view, then the solving process may become very difficult. This suggests the actual written work is an integral part of the cognitive process of solving certain problems. To understand the problem-solving process, it would be useful to investigate the relationships between the nonconscious part, the conscious (mental) part, the external physical part, and the way relevant information is brought to mind. Could there be ways of writing when attempting the solution of a problem that enhance a student's cognition?

These ideas extend far beyond the teaching and learning of elementary number theory. However, number theory would be an ideal setting in which to investigate them.

THEORETICAL FRAMEWORKS

Several monograph authors refer to interesting theoretical frameworks, that is, collections of specialized, interrelated concepts that help express their points of view. One of these, APOS theory, is used by Ferrari (this volume) and is central in the analysis of Brown and colleagues (this volume). Campbell and Zazkis have previously relied on APOS theory and here it becomes part of their view of number theory as a conceptual field, in the sense of Vergnaud. While we agree this can be a fruitful point of view, we also think APOS theory might be more widely applied, and to that end, we will sketch an extended view of it.

APOS is an acronym for action, process, object, and schema, which together form a classification system for ways an individual might conceive of a particular concept. Here, in discussing an idea such as number or function, we will reserve the term *concept* for a building block of "public knowledge"—knowledge in a form that can be communicated, defined, described, and written down. In contrast, we will speak of a *conception* as a (roughly) corresponding building block of an individual's "private" mental knowledge—knowledge that can become conscious, but cannot be precisely communicated or directly observed, only inferred. Because APOS has been described by Brown and colleagues (this volume), we will only recall it through an example. A student, on first hearing about functions, might look at $f(x) = 3x + 2$ and, guided in a procedural way by the (external) written expression, be able to say $f(x) = 5$ when $x = 1$. If he or she could only work with specific examples and could not think about or discuss functions in general, he or she would have an action conception. Repeating this kind of action a number of times and reflecting on it can lead to the ability to think about functions without having a specific function in mind, that is, can lead to interiorizing the experience into a process conception of function. With such a conception, the student could think in a general way about carrying out the required operations without having a specific formula such as $f(x) = 3x + 2$ or might think of input and output, but still might not see this process as a whole or single unit that could be acted upon. On (probably considerable) reflection he or she might encapsulate his or her process conception of function into an object conception, that is, be able to think of it as a single unit that could be acted upon.

While the idea of an object conception of function seems unproblematic—functions can be regarded metaphorically as objects rather like physical objects—Brown and colleagues' "object conception of divisibility" seems counterintuitive because in the everyday sense "divisible" is a property (of two numbers) rather than an object. In the context of APOS theory, there seem to be two main ideas associated with "object." One of these, perhaps intended by those referring to "*cognitive* object," is the idea that an object conception is a mental structure that can be experienced in an undivided way, as a whole without parts. The other is that an object conception is like a physical object—something that not only can be seen as an undivided whole, but can also be acted upon, can be an element of a collection, and so on. Separating these two ideas, wholeness and objectness, provides an extended categorization that we see as potentially useful in studying the development of conceptions.

We propose that concepts, and the associated conceptions, can be separated into *objects* (numbers, groups, topological spaces), *properties* (prime, commutative, compact), and *activities* (factoring, solving, dividing, integrating). *Relations* (less than, divides) can be regarded as properties of two or more objects. In this extended categorization, objects can have properties and be the subject of activities; properties can be associated with objects and activities; and activities can transform objects to yield other objects. Thus "object" as used in APOS theory might be better called *entity*, with its key feature being a unitary or gestalt nature that allows it to take on the role of object, property, or activity. One might, for example, have an action, process, or entity conception of the activity, factoring, or of the relation, divisible. This extended framework allows distinctions between kinds of entities; however, it has no effect on the APOS view of schema and does not alter the central ideas behind APOS theory itself.

There appears to be an additional feature that a conception (either action, process, or entity) may or may not have. This might be called *efficacy* and it permits fluent speech and cognition. That is, an *efficacious* conception can be used in cognition (and speech) without consciously "unpacking," or calling to mind, its defining aspects or the experiences that went into its formation. This may appear to resemble the unitary nature of entity conceptions, but is meant only to refer to the way a conception is used in speech or cognition. The distinction between an efficacious conception and an entity conception can be seen in the first transcript of Brown and colleagues (this volume). There, Adam appears to be fluent with his conception of divisor because he begins calculating at once without having to think about the meaning, despite only having an action or process, not an entity, conception.

Among mathematicians, instances of both this efficacy feature of conceptions and the lack of it are common. For example, most mathematicians have an efficacious, entity conception of the concept "group"; they can understand or speak complex sentences about groups without bringing the definition to mind. In contrast, consider a somewhat obscure mathematical definition from the 1950s: a *clan* is a compact, connected topological semigroup with identity. We suspect that many mathematicians would have to consciously review the meaning of "clan" to fully comprehend, "In any clan there is a minimal subclan connecting the identity to the minimal ideal." That is, they would not have an efficacious conception of "clan" even though they would see it as an object.

On being given an unfamiliar abstract definition, mathematicians seem to treat it logically as an entity, and to suspend or ignore the confusion associated with not having either an efficacious or an entity conception, while they work procedurally at the action or process level, accumulating the experiences needed to yield an efficacious entity conception. Many university students appear unable to apply this "bootstrap" process and may even be unaware that such a thing is possible. They may also avoid use of conceptions that are not efficacious. We suspect that beginning number theory students who are not fluent with one of their conceptions—divides, multiple, or factor—may avoid using it and the symbolic representation it calls for, perhaps substituting more familiar terminology of their own, for example, divides evenly. It would be good to know the degree to which fluency (efficacy) affects use.

Zazkis (this volume) remarks in her conclusion that many preservice elementary teachers are reluctant to adopt formal mathematical terminology, for example, persistently using terms like "goes into" instead of "divides." In explanation, following Sfard, she suggests that mathematical objects emerge in an interaction between metaphor and rigor. The terminology we have introduced here allows us to suggest another explanation.

We suspect that, as children, many of the students Zazkis is referring to constructed a relation conception of divides based on an activity, namely, that of long division with remainder zero. For many, this conception would be efficacious and have the sign (name, symbol) "divides evenly" (or some synonym), with "divides" treated as a contraction of "divides evenly." While for young children such a conception might be metaphorical because numbers might be seen as attached to specific sets, for adults numbers have become objects in their own right and this conception seems neither especially metaphorical nor unrigorous. Later, when the standard mathematical concept of divides is introduced, the students may construct another relation conception, but one based on an object, that is, the defining equation $A = BC$, and be encouraged to use what to them is approximately the same sign, "divides." It is hard to see how two different conceptions with (essentially) the same sign could both be efficacious. Furthermore, it seems unlikely that the first conception ("divides evenly"), or any conception, could be eradicated—short of waiting a very long time. To take an everyday example, can people stop having a conception of car, or for that matter, of unicorn? In this situation, students might best be helped to construct a schema containing both conceptions so that when one comes to mind, the other is also readily available. Fortunately, the two conceptions are complementary. The one based on an activity can yield numerical results such as 15|1080, while the object conception can be used to support reasoning and proofs.

In addition to APOS theory, Brown and colleagues (this volume), introduce a second theoretical framework inspired by the three stages Piaget (1978) employed to describe children's progress from successfully executing a physical task *without* understanding to doing so *with* understanding. Brown and colleagues substitute a procedural mathematical task for the physical task. They follow Piaget in taking indicators of understanding to include the ability to explain why parts of the procedure work and the ability to predict the results of carrying out parts of the procedure before actually doing so. This rather broad perspective on understanding complements that of APOS theory, which has focused mainly on the development of particular conceptions. Furthermore, substituting a procedural mathematical task for a physical task may turn out not to be a very radical change. Much procedural work in mathematics appears to be partly physical in nature, that is, to depend on the partly physical acts of writing and reading in a way inextricably integrated with the mental activity. For example, in solving an equation, one of our calculus students wrote "+ 4" so that the horizontal bar of the "+" was half complete, causing the inscription to resemble "− 14," after which the student erroneously used "− 14." Such errors depend specifically on interpreting the most recent inscription and could not happen if the written work was merely a (partial) transcription of otherwise wholly mental activity. Writing and seeing what she had written seems to have been an essential part of this student's cognition.

Using the three-stage framework to examine the way procedural knowledge supports understanding in elementary number theory allows Brown and colleagues to give an interesting picture of their preservice teachers' knowledge. But it also suggests a much broader application. The teaching and learning of procedural knowledge has been neglected in recent years by the mathematics education research community (Star, 2000). Yet procedural knowledge is often most of what is assessed at the lower division undergraduate level and some mathematicians believe it supports, and its mastery must precede, the teaching of the corresponding conceptual understanding. While there is evidence this last belief is inaccurate, that the mastery of procedures need not precede the teaching of conceptual understanding (Heid, 1988), procedural knowledge does appear to support the construction of conceptual understanding in two main ways. It is needed to provide situations, patterns, and experiences upon which reflection can lead to the construction of new conceptions and understandings. By way of illustration, one can learn to execute the rules of the game of checkers (procedurally), after which one has an opportunity to reflect and discern its strategies (Simon, Tzur, Heinz, & Kinzel, 2000). In addition, some understanding appears to be constructed out of reflection on procedures themselves—the basis of the three-stage framework. Perhaps one might think of a procedure as providing the central core of one kind of explanatory schema.

Because of the nature of their data—single interviews with preservice teachers—Brown and colleagues mainly describe the current state of their subjects' knowledge, not lasting changes in knowledge that would indicate learning. However, in a number of transcripts, one can see students actively constructing understanding, for example, Alice progresses from Stage I to Stage II and Karyl moves through Stage II to Stage III. This suggests single student teaching experiments consisting of sequences of probing interviews might be designed specifically to encourage the construction and retention of new understandings. By focusing as much on the interviewer as the student, it might be possible to identify situations that facilitate students' progress from Stage I to Stage III, that is, that facilitate the construction of conceptual understanding from procedural knowledge. Elementary number theory would be a good topic for such a study because it can be taught early, minimizing the effects of prior knowledge and habits of mind so as to isolate situations that facilitate conceptual learning. Any results, however, might be of interest far beyond the teaching of elementary number theory.

IMPLICATIONS FOR TEACHING

This monograph contains a number of mathematically rich number-theoretic tasks. With them, teachers could try to probe students' understanding or help them with generalization and abstraction. It seems that through the use of an unexpected representation, students can be disequilibrated and sometimes brought to reconsider their knowledge. Asking what would be the remainder when $M = 3^3 \times 5^2 \times 7$ is divided by 15 (Campbell, this volume) or whether $M = 3^4 \cdot 5^3 \cdot 7^6 \cdot 19^8$ is divisible by 63 (Ferrari, Chapter 5, this volume) seem sufficiently novel that students first resort to their calculators, and when pushed (or finding the calculator of no use), try other approaches.

The two problems: (1) Can you find a rule for determining the number of different divisors of any number, given its prime factorization? (Teppo, Chapter 6, this volume); and (2) given an $N \times K$ grid, how many squares are crossed by the diagonal? (Edwards & Zazkis, this volume), are requests for a generalization, with the accompanying need to consider examples and make and test conjectures. Both problems can be used with preservice elementary teachers as a way of introducing them to these important aspects of mathematical reasoning. In a somewhat similar way, the six process pattern generalization problems, for example, find an upper bound for the sequence $\sqrt{2}, \sqrt{2+\sqrt{2}}\ \sqrt{2+\sqrt{2+\sqrt{2}}}, \ldots$ (Harel, this volume), can prod at least some students to find relationships between consecutive terms and explain their reasoning (in the form of quasi-induction) as a precursor to formal mathematical induction.

In addition to such tasks, ideas/principles for teaching mathematics can be drawn from the chapters, although some of these can also be found elsewhere. Teppo's classroom vignette can serve as a prototype for guiding students to think for themselves, using appropriate teacher scaffolding and a suitable didactic object (divisor table), while also leading to a more general question about numbers of divisors.

The interviews of Brown and colleagues (this volume) were not meant as teaching sessions, but seemed to help students develop insights. Perhaps with some modest adjustments they could serve as a model for teaching the understanding of procedures on an individual or small group level. Providing individual assistance is a small but important role of many university mathematics teachers. The semi-structured nature of the interviews, with lists of problems and questions, might prove helpful to teachers. Indeed, it might be useful to circulate such interview material, together with annotations on student reactions, among teachers as a form of pedagogical content knowledge.

The value of appropriate generic proofs for exposition in number theory was argued convincingly, along with some general principles for selecting a particular case. There is also the admonition that writing up a more formal proof can be a nontrivial task for students (Rowland, this volume).

The idea that the type and sequence of problems is of "paramount pedagogical importance" is not new, nor is the idea that such sequencing of tasks should follow students' conceptual development (Harel, this volume). What is new is the somewhat counterintuitive notion that teaching, at least for mathematical induction, should begin with suitably chosen "harder" problems, rather than the more usual progression from "easy" to "hard," as preconceived by textbook authors. Other factors that probably influenced students' success in the mathematical induction teaching experiment were the opportunity for repeated reasoning of an appropriate type (process pattern generalization), the opportunity to clarify and rewrite one's solutions, and the request that students summarize the most important ideas of the mathematical induction unit in their own language, followed by the instructor's incorporation of these into a presentation in precise mathematical language.

Elementary number theory tasks are a rich and accessible area with which to enhance and probe students' mathematical sense making. Many fascinating research directions/questions have been considered in this monograph. Some are specific to elementary number theory, but others suggest using number theory tasks to investigate more general aspects of mathematical learning and cognition.

REFERENCES

Bills, L. (1996). The use of examples in the teaching and learning of mathematics. In L. Puig & A. Gutierrez (Eds.), *Proceedings of the 20th Conference of the International Group for the Psychology of Mathematics Education* (Vol. 2, pp. 81–88). Valencia, Spain: University of Valencia.

Billstein, R., Libeskind, S., & Lott, J. (1984). *A problem solving approach to mathematics for elementary school teachers* (2nd ed.) Menlo Park, CA: Benjamin Cummings.

Bolker, E. D. (1970). *Elementary number theory: An algebraic approach.* New York: W. A. Benjamin.

Epp, S. S. (1990). *Discrete mathematics with applications.* Belmont, CA: Wadsworth.

Grouws, D. A. (Ed.). (1992). *NCTM Handbook of research in mathematics teaching and learning*. New York: Macmillan.

Hazzan, O. (1999). Reducing abstraction level when learning abstract algebra concepts. *Educational Studies in Mathematics, 40*(1), 71–90.

Heid, M. K. (1988). Resequencing skills and concepts in applied calculus using the computer as a tool. *Journal for Research in Mathematics Education, 19*, 3–25.

James, W. (1910). *The principles of psychology*. New York: Holt.

National Council of Teachers of Mathematics (2000). *Principles and standards for school mathematics*. Reston, VA: Author.

Peterson, I. (2000). The power of partitions: Writing a whole number as the sum of smaller numbers springs a mathematical surprise. *Science News, 157*(25), 396–397.

Piaget, J. (1978). *Success and understanding*. Cambridge, MA: Harvard University Press.

Pirie, S. E. B. & Kieren, T. E. (1994). Growth in mathematical understanding: How can we characterise and how can we represent it? *Educational Studies in Mathematics, 26*, 165–190.

Simon, M. A., Tzur, R., Heinz, K., & Kinzel, M. (2000). Articulating theoretical constructs for mathematical reasoning. In M. L. Fernández (Ed.), *Proceedings of the 22nd Annual Meeting of the North American Chapter of the International Group for the Psychology of Mathematics Education* (Vol. 1, pp. 213–218). Columbus, OH: ERIC Clearinghouse for Science, Mathematics, and Environmental Education.

Star, J. R. (2000). Re-"conceptualizing" procedural knowledge in mathematics. In M. L. Fernández (Ed.), *Proceedings of the 22nd Annual Meeting of the North American Chapter of the International Group for the Psychology of Mathematics Education* (Vol. 1, pp. 219–223). Columbus, OH: ERIC Clearinghouse for Science, Mathematics, and Environmental Education.

Stiff, L. V. (1999). *Developing mathematical reasoning in grades K–12: 1999 yearbook*. Reston, VA: National Council of Teachers of Mathematics.

Tomasello, M. (1997). Human see, human do. *Natural History, 106*(8), 45–47.

Tomasello, M. (1999). *The cultural origins of human cognition*. Cambridge, MA: Harvard University Press.

Towers, J., Martin, L., & Pirie, S. (2000). Growing mathematical understanding: Layered observations. In M. L. Fernández (Ed.), *Proceedings of the 22nd Annual Meeting of the North American Chapter of the International Group for the Psychology of Mathematics Education* (Vol. 1, pp. 225–230). Columbus, OH: ERIC Clearinghouse for Science, Mathematics, and Environmental Education.

Author Index

Adler, J., 94
Armstrong, B. E., 118
Asiala, M., 43, 44

Badger, M., 159
Baker, A., 169
Balacheff, N., 160
Ball, D., 15, 32
Baxandall, P. R., 193
Beardon, A., 179
Bell, E. T. (1883–1960), 163
Bills, L., 172, 214
Billstein, R., 217
Blume, G., 141
Bolker, E. D., 216
Bouffi, A., 126
Boyer, C., 185
Brewer, W. F., 142
Brousseau, G., 208
Brown, A., 9, 106, 132, 218, 219, 224, 225, 227, 228, 229
Brown, J. S., 140
Brown, W. S., 193
Burn, R. P., 180 n.8
Buschman, L., 83

Campbell, S. R., 7, 9, 15–18, 21, 32, 34, 42, 44, 45, 47, 49, 50, 68, 78, 83, 85, 87, 89, 91, 92, 106, 108, 113, 124, 132, 218–21, 224, 228
Chazan, D., 140, 142
Chiappini, G., 104
Chinn, C. A., 142
Clark, H., 99
Clark, J. M., 44, 47
Cobb, P., 83, 126, 188, 240, 241
Collins, A., 140
Collis, K. F., 166
Copes, L., 164
Cramer, K., 83

Dautermann, J., 106
Davis, P. J., 141
Deri, M., 16
Dörfler, W., 6
Drouhard, J.P., 108

Dubinsky, E., 7, 44, 106, 113, 185, 188, 193
Duguid, P., 140
Durkin, K., 83, 87
Duval, R., 99, 100, 114

Edwards, L. D., 11, 141, 142, 228
Epp, S. S., 217, 218
Ernest, P., 185, 191, 193
Esty, W. W., 83

Fauvel, J., 162
Feghali, I., 15
Fennema, E., 117
Ferrari, P. L., 10, 100, 224, 228
Fischbein, E. (1920–1998), 16, 36, 38
Flores, A., 117
Foxman, D., 159
Freudenthal, H. (1905–1990), 12, 16, 38, 41, 46

Gadowsky, K., 132
Glasersfeld, E. v., 5, 6, 210 n.5
Graham, R., 163
Greer, B., 1, 33, 211 n.6
Grice, H. P., 99
Grouws, D. A., 213

Hanna, G., 140, 158
Hanselman, C. A., 83
Hardy, G. H. (1877–1947), 2
Harel, G., 11, 140, 141, 162, 180 nn.1, 6, 186, 192, 193, 201, 206, 207, 210 n.1, 211 nn.6, 14, 216, 228, 229
Hazzan, O., 215
Healy, L., 4, 83
Heid, M. K., 227
Heinz, K., 227
Hersh, R., 141, 158, 159, 164, 179, 180 n.1
Hewitt, D., 178
Hiebert, J., 33
Hoffman, P., 163
Hoyles, C., 4, 83

James, W. (1842–1910), 220

Kant, I. (1724–1804), 5, 6
Kaput, J. J., 106
Karnowski, L., 83
Kieran, C., 2
Kieren, T. E., 218
Kinzel, M., 227
Klein, J. (1899–1978), 12
Kornblith, H., 5

Lave, J., 7, 94, 140
Leonard, F., 33
Leron, U., 106, 177, 181 n.14
Libeskind, S., 217
Loomis, E. S., 180 n.3
Lott, J., 217

Ma, L., 2, 57, 79, 80
Mack, N., 1, 7, 33
Magone, M., 33
Maher, C, 140
Manaster, A., 201, 210 n.1
Marino, M. S., 16
Markovits, Z., 33
Martin, L., 218
Martin,W. G., 141
Martini, R., 159
Martino, A., 140
Mason, J., 83, 141, 160, 168, 178, 181 n.16
Mathematical Sciences Education Board, 2
Matz, M., 22, 24
McClain, K., 126
McClure, J., 201, 210 n.1
McDonald, M., 44
Molinari, M., 104
Morgan, C., 99
Morley, A., 161

Nello, M. S., 16
Nelson, R. B., 162
Nesher, P., 33, 83
Newell, A., 8

Omanson, S., 33
Ono, K., 214

Peled, I., 33
Pence, B., 141
Peterson, I., 214
Philipp, R. A., 117, 118
Piaget, J. (1896–1980), 5–8, 41–43, 45–47, 72, 106, 210 n.5, 222, 227
Pimm, D., 87, 99, 140, 141, 160, 168, 178
Pirie, S. E. B., 218
Polya, G., 162

Resnick, L. B., 33
Romberg, T. A., 117
Rose, G. St. C., 193
Rowland, T., 11, 161, 168, 177, 214, 216, 229
Ruddock, G., 159
Russell, B. (1872–1970), 8

Schappelle, B. P., 117, 118
Semadeni, Z., 161, 167, 168, 177, 179
Sfard, A., 6, 83, 93, 94, 226
Shire, B., 83, 87
Show, B., 191
Sierpinska, A., 44
Silver, E. A., 1
Simon, H. A., 8
Simon, M. A., 15, 32, 35, 38, 141, 227
Simonsen, L. M., 118
Singh, S., 159
Sowder, J. T., 33, 117, 118
Sowder, L., 140, 141, 162, 180 n.1, 186, 192, 193, 201, 210 n.1, 211 n.8
Star, J. R., 227
Steffe, L., 188
Streefland, L., 83
Stiff, L. V., 213
Sutherland, R., 83

Tall, D., 201
Teppo, A. R., 10, 83, 118, 214, 228
Teubal, E., 83
Thomas, K., 9, 106, 132, 218
Thompson, P. W., 126
Tolias, G., 9, 106, 132, 218
Tomasello, M., 221
Towers, J., 218
Tzur, R., 227

Vergnaud, G., 6–9, 17, 199
Vinner, S., 97, 189
Vuyk, R., 210 n.5
Vygotsky, L, 8

Walther, G., 161, 163, 167, 168, 173
Wason, P.C., 142
Watson, F.R., 193
Wearne, D. C., 33
Wenger, E., 94
Wheeler, M. M., 15
Whitenack, J., 126
Wilcs, A., 159, 180 n.2
Woodall, D., 193

Yackel, E., 140, 141

Zan, R., 98
Zazkis, R., 7, 10, 11, 15–18, 21, 32, 34, 37, 42, 44, 45, 47, 49, 50, 51, 68, 78, 83, 84, 85, 89, 91, 93, 106, 108, 113, 124, 128, 132, 152, 218, 220, 221, 224, 226

Subject Index

algebra, 44, 97, 203, 223; abstract, 44, 97, 116, 215; learning and teaching, 98, 215–16, 223; relation to number theory, 1, 3–5, 8, 12, 108, 116, 123
algebraic: argument and proofs, 172, 215; computation and error, 24, 194; expressions, language, symbolism, and formalism, 98, 100, 122, 123, 125, 128, 152, 158, 192; function and structure, 99, 127
APOS (Action-Process-Object-Schema) theory, 7, 10, 41–45, 51, 76–80, 106–8, 113, 116, 224–28; action, 6, 7, 9–10, 41–48, 51–62, 65–73, 75–81, 106–7, 109, 111, 113, 218, 220, 224–26 (*See also* generic proof; reflection); action conception, 43–44, 224; deencapsulation, 43, 60–61, 67–68; disequilibration and reequilibration, 24, 55, 77, 106, 217, 222, 228; encapsulation, 7, 43, 45, 67–68, 78–79, 106, 107, 116, 224–25; extensions, 6–8, 47–48, 106–7, 224–28; genetic decomposition, 44–45; interiorization, 43, 45, 52, 77, 106–7, 110, 224; limitations of, 7; object, 6, 7, 43–45, 51, 67, 68, 71, 76, 77–79, 106, 107, 109, 110, 225, 226 (*See also* mathematical objects); object conception, 45, 64, 77, 225, 227; process, 7, 24, 42–46, 50–51, 52, 53, 55–57, 59–64, 66–68, 71, 75–79, 86, 89–91, 100, 106, 107, 109, 110, 112, 113, 116, 136, 222, 224–26; process conception, 43–45, 55, 62, 64, 70, 71, 224, 225; schema, 6, 7, 43–48, 51, 57, 62, 65, 68, 73, 74, 76, 77, 80, 81, 82, 100, 105, 106, 113, 224–26 (*See also* proof scheme); thematization, 7, 10, 21, 37, 44, 80, 81 n.2, 106, 107, 111, 112
arithmetic, 1, 2, 3, 4, 7, 8, 11; actions, activities, ideas, notions, concepts, understanding, representation, thinking, and reasoning, 1, 4, 7, 8, 11, 12, 20, 29, 32, 33, 34, 41, 47, 61, 79, 80, 107, 123, 124; fundamental theorem of, 8–9, 68, 79, 101, 124, 134 (*See also* prime decomposition); mental, 176;

modular, 17, 38, 163, 165, 166–67, 169–70, 174, 180–81 n.9, 181 nn.12, 13, 188, 220; properties, operations, and procedures, 1, 7, 11, 32, 34, 41, 42, 44, 46, 57, 63, 76, 80, 92, 110, 135; rational vs. integer (whole) number, 1, 12, 17, 30–32, 220; relation to number theory, 1–5, 8, 12; sequences, 4, 51, 79, 93, 131–36, 166, 190–92, 194, 195–97, 200, 203, 215, 216, 228; units (divisible vs. indivisible), 12, 20, 36–37, 38–39, 220; well-formed formula (equation), 28–29, 35, 37

behaviorism, 221
Boole, G. (1815–1864), 223
Boolean algebra, 223

Chinese Remainder Theorem, 163, 180–81 n.9
classroom discourse and practice, 9, 10, 80, 83, 117, 159, 170, 188, 202, 203, 205, 228. *See also* reflective discourse
classroom investigations, 117–28, 170–75, 188–205; complexity of, 121, 125, 128
cognitive content and processes: emerging from social interaction (acculturation), 140; integration of, 117. *See* APOS theory; abstraction; conjectures; generalization, knowledge; mathematical object; mathematical process; reasoning; reflection; representation; understanding
cognitive modeling, 6–8. *See also* APOS theory; conceptual field
cognitive structure, 7, 47. *See also* APOS theory; conceptual field
communication, 9, 83, 87, 93, 96, 98, 124–26, 128, 196. *See also* reflective discourse
concept formation, 6–7, 209. *See also* APOS theory; conceptual development
concept (mental) image, 124, 189–90, 219–20
conceptual development, 195–203, 229. *See also* APOS theory; conceptual field
conceptual field, 6–9; number theory as a conceptual field, 8–11
conceptual knowledge and understanding. *See* knowledge; understanding
congruence, 3, 108, 165–67, 180–81 n.9, 181 n.13, 215
conjectures, 9, 10, 11, 89, 118, 118–23, 139–54, 157, 159–60, 163, 165, 174, 176, 192, 214 (*See also* proof; reasoning); closure, 166, 177; cognitive strain, 177; from conjecture to proof, 152–54, 215 (*See also* generalization; proof); counter-examples (disconfirming evidence, refutation), 11, 124, 142–45, 150–51, 174; explanation and justification (conviction), 11, 89, 154, 159–60; insight, 159, 161; making and testing, 10, 11, 118, 120–23, 140–41, 143, 149, 150, 159, 228; naive empiricism, 152; strategies, numerical pattern-seeking, 148–49; strategies, visual decomposition, 147–48; validity, 141–42
constructivism, 5–6, 220–24; approximate knowledge and truth, 188, 222; epistemological tradition, 5; instructional approach, 48; objectivity and subjectivity, 5, 188, 221–22; radical, 5–6, 221–23; schema, 6 (*See* APOS theory; conceptual fields); uncertainty (probabilistic and synthetic), 222
content knowledge. *See* knowledge

Descartes, R. (1596–1650), 5
didactic object, 126–27, 128, 214, 228
distributivity, 30, 34, 41, 46
divisibility, 3, 7–11, 15–18, 21, 23, 26, 28, 31–38, 41–82, 84–94, 98, 107, 109–11, 113, 114, 118, 121, 124, 132, 136, 143, 152, 153, 162, 176, 188, 190, 194, 213, 214, 216–20, 225; connection with, divides (evenly), 16, 17, 118, 217, 226; connection with, division, 16, 17, 22–24, 37; connection with, encapsulation, 45; connection with, factoring, 23; connection with, fractional component, 16–17, 24; connection with, multiplication, 17, 27;

equivalent statements of, 84, 217; prime factorization, 57
divisibility rules (criteria): 3, 8–9, 18, 21, 108; divisibility by nine, 162, 176
division, 8, 9, 11, 15–39, 41, 42, 45–48, 51–63, 77, 78, 84, 87–93, 110–14, 118, 124, 134, 217–20, 226 (*See also* division with remainder); closure, 36; difficulties with, 36–37; "ideal" understanding of, 37; inverse (reverse) of multiplication, 30–32, 46, 47, 55, 59, 61, 77, 78; long, 15–17, 20–22, 24, 25, 30, 32, 37, 60, 219, 226; meaning of, 17; partitive (sharing), 16, 20–21, 23, 35–36, 37–38, 218; quotitive (measurement), 16, 21, 35–36, 37–38, 90, 218, 220; "taking out" (repeated subtraction), 124, 219; visual images of, 218, 219
division algorithm/theorem. *See* division with remainder
division with remainder (whole number division), 9, 15, 18, 21, 25–32, 34–37, 217–18, 226; definition, 19–20, 32, 34, 37; distinguished from rational number division, 9, 16, 17, 24–26, 34–35, 36, 87, 92; inverse of multiplication (and division), 24, 30–32, 61; misleading notation, 28–30, 34–35, 37; novel and unfamiliar (disequilibrating abstract) contexts, 6, 17–21, 24, 32–33, 34, 36; procedural orientation toward, 21, 28, 30, 31, 32; quotient, 20–21, 22, 23–24, 28–29, 30, 34, 35–36, 37 (*See also* quotient); remainder, 20–22, 24, 25–26, 27, 30, 33–34, 35, 36, 37–38, 219 (*See also* remainder); using calculators, 16–17, 24–28, 30, 33, 37
divisor, 3, 8, 10, 11, 20–23, 25–26, 28–30, 32–39, 44, 46, 49, 51–56, 59–60, 66, 77, 78, 84, 89, 90, 107–9, 111, 113, 118–28, 146–47, 164–66, 170, 172, 176, 188, 214, 217, 225, 228; GCD (Greatest Common Divisor), 3, 8, 98, 107, 108, 110, 146–50, 152, 164, 188, 208; least prime, 166
DNR (Duality, Necessity, and Repeated-reasoning), 186, 205–10, 216 (*See also* proof scheme); duality (thinking and understanding), 207; necessity (intellectual needs), 207–8; repeated-reasoning (internalize and interiorize), 197, 209–10

epistemology. *See* behaviorism; constructivism; logical positivism
Euclid (c.365–c.275 B.C.), 165
Euclidean algorithm, 5, 8; Euclidean division, 107, 113 (*See also* division with remainder); proof of infinite primes, 165–66, 168
Euler, L. (1707–1783), 143, 163, 164, 166, 173
Euler's Theorem, 164; Criterion, 166; ϕ-function, 143, 163–64, 173–74, 216

factor, 3, 8, 10, 11, 21, 29, 45, 56, 58, 60, 61, 64–70, 74, 75, 77, 78, 84–93, 109, 111–13, 118–25, 128, 131, 132, 136, 143, 147–53, 160, 164, 167–68, 176, 204, 217, 219, 226, 229 (*See also* divisors; prime decomposition); common, 75, 110, 133, 136, 147–148, 150, 152, 153, 160; composite, 79; greatest common (*See* divisor, GCD); prime, 61, 67, 68, 70, 109, 110, 120, 123–25, 127, 128, 131, 168; tree, 23, 133
factoring, 21, 23, 109, 225
Fermat, P. (1601–1665), 140, 185
Fermat's Last Theorem, 140, 159, 180 n.2
Fibonnaci (1175–c.1250), 195
Fibonacci numbers, 195
focus, 220–21
function, 1, 3, 12, 44, 98–99, 101, 102, 123, 128, 133, 135, 163–64, 173–74, 177, 190, 199, 207, 215, 216, 224–25; concept (idea) of, 44, 108; derivative of, 207; recursive representation, 190

Gauss, C. F. (1777–1855), 2, 162–163, 166
Gauss's Lemma, 166–67, 176
Gauss's method, 162–63
generalization, 11, 107, 118, 121, 127, 139–43, 191–94 (*See also* conjectures;

proof scheme); empirical, 172, 192, 216; general and particular, 126; general terms, 135, 136; generic examples (*See* generic example; generic proof); making and expressing, 50, 51–55, 70, 118, 121, 122, 141, 153, 206; numerical to algebraic, 123; over-generalization, 35, 36, 47–48, 64; pattern, 119–22, 128; pattern, process, 191–93, 194, 195–201, 203, 206, 216, 228; pattern, result, 191–93, 194, 206; solutions to problems, 139–40; rules and formulae, 119–122, 128, 141
generic example, 11, 157, 160–64, 168–71, 174, 175–80, 216, 220; Balacheff's characterization, 160–61; epistemological purpose, 164; pedagogical deployment, 175–80
generic proof, 157–83, 213, 216, 229; "action proof," 161–62, 169; general (guiding) principles, 167–69, 229; "illuminating example," 161, 162; intuition, 180; invariants, 168, 173; limitations, 165; transfer, 173
Goldbach, C. (1690–1764), 140
Goldbach's conjecture, 140

infinite, 64, 107, 165, 168, 189, 200, 220
insight, 21, 77, 110, 159–60, 161, 168, 176, 177, 229

Jailer problem, 160, 176–77

Kant, I. (1724–1804), 5, 6
knowledge (*See also* understanding): approximate, 222–23; conceptual, 220; and constructivism, 6, 220–23; content, 15, 18, 32, 33, 78, 80, 97, 118, 229; existing, implicit, informal, local, and prior, 7, 17, 47, 102, 103, 108, 125, 175, 199, 208, 228 (*See also* theorem-in-action); mathematical, 6, 43, 49, 78, 79–80, 98, 164; of multiplicative structure and divisibility, 41, 42, 48, 50, 51, 61, 65, 66, 75, 76, 78, 80–81; by osmosis, 158; objective vs. subjective (public vs. private), 5, 224; procedural, 221, 227–28; procedural, imitation and emulation, 221; student (learners' constructed), 42, 47, 99, 102–3, 204, 209, 218, 220–22, 227–28; zone of proximal, 200

Lagrange, J. L., (1736–1813), 181 n.18
Lagrange's Theorem, 181 n.18
language (*See also* classroom discourse and practice; communication; learning; meaning; representation): ambiguity (need for precision and lack of mastery), 10, 17, 33, 34–35, 37, 83, 89, 98, 103, 110, 116, 167, 229; analysis of, 108, 124; context, 1, 7, 21–24, 124, 140; implicature, 99; metaphorical, 90, 141, 225; natural (normal, ordinary, informal) and mathematical (algebraic, symbolic, formal), 1, 7, 10, 35, 36, 83–94, 98, 99–100, 125, 128, 158, 196, 202, 229; of number theory, 32–34, 83–94, 217–20; register (natural and formal), 10, 87, 99, 100; semiotic control, 10, 97–115; vocabulary (categorization, classification, terms and expressions), 5–6, 16, 19, 33, 36, 38, 60, 79, 84–94, 98, 99–101, 104–5, 108, 110, 124–25, 196–97, 217
learning. *See* APOS theory; conceptual development; knowledge; learning difficulties; learning strategies; problem contexts; problem solving; problem tasks; understanding; reasoning
learning difficulties, errors, misconceptions, and mistakes (*See also* communication; language): ascribed to educational practice, 103, 136, 169–70, 189, 194–95, 206; awareness (or lack) of, 11, 135, 150, 159, 224; with calculations, procedures, and techniques, 1, 12, 17, 21, 24–25, 28–29, 32, 33, 35, 109, 158, 175, 191, 227; with concepts, ideas, and understanding, 2, 12, 15, 16, 17, 21, 24, 28, 30, 32, 33–34, 38, 60, 69, 92, 97–98, 117, 119, 132, 152, 153, 168, 189, 190, 206, 215, 219–20; with language, meaning, and representation, 1, 12, 16, 17, 33, 38, 42, 54, 85, 92, 98, 102, 105, 107, 110–11,

113, 116, 132, 219–20; with reasoning, 158, 177, 189, 206
learning strategies (and techniques), 9, 10, 11, 50, 53, 57, 71, 79, 84, 90, 98, 100–103, 109, 111, 112–13, 116, 131, 133–35, 141, 146–49, 157, 158, 185, 189, 191, 205, 227; pedagogical (teaching), 84, 157; problem solving, 11, 98, 133–35, 146–49; proofs and conjectures (*See also* proof; conjectures), 9, 11, 141, 147–49, 157, 158, 160, 185, 189, 191, 205; "symbol pushing," 101; "trial and error," 48, 66, 69–73, 80, 98, 102, 122, 123
logarithms, 98, 191–93, 198, 204

mathematical, meaning, 1, 6–7, 9, 10, 11, 16, 17, 21, 23, 24, 32, 34–35, 38, 50, 51, 52–53, 54, 55, 87, 91–93, 98, 99, 100–101, 102, 103, 135, 193 (*See also* language; learning; understanding); modeling, 12; objects (*See* mathematical objects); operations, 1, 2, 4, 5, 7–9, 11, 16, 24, 34, 36, 41, 43–47, 57, 63, 64, 76, 77, 79–81, 83, 87, 123–25, 128, 135, 161, 168, 170, 199, 215, 223, 224; procedures, 1, 5, 7–10, 15–19, 21, 25, 31, 32, 36, 45, 51, 54, 65, 67, 68, 70, 72–73, 75, 77–80, 100, 102–4, 107, 110, 117, 118, 123–24, 126, 139, 158, 193, 218, 227, 229; processes, 9, 10, 24, 48, 49, 53, 66, 86, 89, 91, 97, 100, 117–19, 121–23, 125, 126, 128, 136, 139, 141, 158–60, 162, 170, 172, 174, 178, 186, 188 (*See also* APOS theory, process; quasi-induction); properties, 1, 3, 4, 7, 8, 11, 12, 33, 41, 42, 45- 47, 55, 57, 68, 69, 75–79, 98, 100, 101, 102, 106, 107, 109–112, 116, 123, 131, 134, 152, 160, 164, 168, 185, 217, 219, 223, 225; structures, 1, 4, 7–12, 16, 33, 38, 41–43, 46–48, 50, 51, 56, 57, 59, 61, 62, 65–68, 71, 73–74, 76–81, 87, 93, 118, 121-25, 127, 128, 133–35, 152, 158, 176, 200–2, 205, 209, 213 (*See also* APOS theory, schema); thinking, 1, 32, 207 (*See also* NCTM standards)
mathematical objects, 6, 89, 93, 125, 168, 176, 177, 180 n.6, 215, 216, 219 (*See also* APOS theory, object); of discourse, 125, 126, 226 (*See also* didactic object); existence and uniqueness of, 109–110, 166, 208; invariance, 168, 176 (*See also* generic examples); relations between, 168, 176, 180, 225; as unknowns, 208
metaphor, 83, 93–94, 215, 226. *See also* language
methodology, 3, 6–8, 17–19, 47, 48–50, 143, 188–89
MI (Mathematical Induction), 9, 11, 44, 158, 185–212, 213, 216, 228, 229 (*See also* conjectures; generalization; proof scheme; quasi-induction; reasoning); abstraction of quasi-induction, 201–3; domino analogy, 189, 211 n.6; formal principle of, 201–3, 206; historical development, 201, 206; internalization and interiorization, 195, 206, 209, 210, 211 n.9, 216; non-recursion problems, 190; recursion problems, 190, 194–201, 207; standard instructional approach, 190–95; undergraduate understandings of, 152, 158, 190, 194–205, 206
multiple, 3, 8, 10, 11, 17, 21, 46, 49–54, 58, 62–75, 77–80, 84, 86, 88, 90, 103, 108, 111–13, 132–36, 143, 149, 152, 153, 160, 165, 167, 173, 176, 214, 217, 218, 226; common, 66, 68–71, 73, 74, 77; consecutive, 51, 55, 132, 133, 135; higher exponent rule, 68, 79, 132, 136; LCM (Least Common Multiple), 3, 8, 50, 65–70, 73–75, 77, 79, 80, 132, 136; repeated addition, 51, 62–64, 217
multiplication, 11, 19, 24, 30–32, 41, 42, 45–47, 51, 53–57, 59, 61–64, 66, 68, 69, 71, 77–79, 118, 123, 124, 133–36, 165, 170, 215, 217–18
multiplicative structure. *See* APOS theory; mathematical, structure

NCTM (National Council of Teachers of Mathematics) standards, 2, 4–5, 15, 19,

83, 117–18, 125, 128, 131–32, 139, 213–14
number, rational, 4, 11, 12, 20–21, 24–26, 33–34, 102–104, 118; rational, decimal, 25, 33, 91–92, 102–103, 112; rational, fractions, 35–38, 54, 92, 93, 103–4, 107, 112, 113, 118, 200, 220; rational, proportions, 118, 149; rational, ratios, 118, 135, 148, 151; whole (integer), even and odd, 30, 109, 111, 113, 148–54, 162, 174; whole, patterns and sequences, 118–28, 131–36; whole, prime and composite, 2, 8, 55, 79, 85, 118, 120–22, 124, 168, 181 n.13 (*See also* prime numbers)
number theory: advanced topics, 3, 9; algebraic connections, 4, 123, 125; as a conceptual field, 8–11; elementary (introductory) topics, 3, 9; as higher arithmetic, 2; historical development, 1–2, 8, 12, 201, 206; and the NCTM curriculum and content standards (*See* NCTM standards); perennial allure, 2

Oxford University, 181 n.21

partition(s), 20, 147, 159–60, 214–15
pedagogical challenges (suggestions and opportunities): divisibility, and prime decomposition, 98–99, 136; division and multiplication, 35–36, 45, 78–80; divisors and multiples, 51, 118–19 125–27, 136; generic examples, 175–80; language and communication, 83, 93–94, 126; proof schemes, 186, 205–10
pedagogy and instruction: assessment and design, 44; cautions, deficiencies, goals, and expectations, 34, 83, 117–18, 124–25, 131–32, 139–40, 178, 188, 194–95, 213–17; challenges (*See* pedagogical challenges); implications, 2, 228–29; instructional process, 9, 10, 117, 119, 210; problem solving (*See* problem solving); working in groups, 48–49, 196, 200
Piaget, J. (1896–1980), 5–8, 41–43, 45–47, 72, 106, 210 n.5, 222, 227
Plato (430–349 B.C.), 177
prime decomposition (prime factorization), 8, 15, 21, 101, 109, 127, 132, 213, 214, 217, 218, 219, 228 (*See also* factors, prime numbers); "breaking down" (factor trees), 23, 67; and divisibility, 42, 55–62, 77; infinity of (*See* Euclid, proof of infinite primes); and (least common) multiples, 62–79; and sequences, 131–136; understanding division, 21–24; uniqueness. (*See* arithmetic, fundamental theorem of)
prime numbers, 118, 120–22, 165–66, 168, 171, 173, 174–75; primitive roots, 169–170, 171, 178, 181 n.11; relatively (co-) prime, 3, 132, 146–50, 163–64, 167, 173–74, 180 n.7, 180–81 n.9, 181 nn.10, 13
problem context, 7, 18–24, 32–34, 36, 41, 43, 46–48, 64, 65, 66, 72, 87, 103, 108, 134, 158–159, 180 n.5, 185, 202–203, 205; novel, unfamiliar, abstract, mathematical, 18, 19, 20, 21–24, 32, 33, 34, 36, 41, 87, 134, 185, 202–203, 205; situated, physical, familiar, limited, concrete, everyday, social, 7, 32, 43, 46–48, 64, 65, 140, 145, 202
problem solving, 2, 12, 42, 97–98, 100, 108, 128, 140, 141, 150, 158, 180, 194–205, 214–17, 221 (*See also* learning); conscious vs. unconscious, 223–24; strategies, 146, 198. *See also* conjectures, quasi-induction
problem tasks, 16, 18–19, 49–50, 101, 104, 108–109, 118–19, 131, 136, 144, 195, 197–98, 199 204. *See also* learning
procedural knowledge and understanding. *See* knowledge; understanding
proof (*See also* conjectures; generalization; generic proof; proof scheme): constructive, 159–60, 198; deductive, 158; deductive argument, 159; *diknumi*, 162; indirect (nonconstructive, by contradiction), 109, 166, 177, 199; induction, 185–211 (*See also* MI; quasi-induction); purposes of, 158–59
proof scheme, 162, 180 n.6, 185–211; conceptual framework, 187; deductive,

180 n.6, 186, 192–93, 194, 202, 203–5, 206, 207; definition, 186–88; empirical, 186, 190, 191–93, 194, 206, 207; external, 186, 190, 193–94, 206, 207; mathematical induction (*See* MI); pedagogical principles (*See* DNR); processes (ascertaining and persuading), 180 n.1, 186, 201, 210 n.5

preservice elementary teachers: knowledge (*See* knowledge, content); understanding of, arithmetic in abstract contexts, 36; understanding of, conjectures, 117–28, 139–54; understanding of, division, 15–40; understanding of, divisibility and multiplicative structure, 41–82; understanding of, language of number theory, 83–96 (*See also* language)

professional development, 34, 117–18, 131–32, 139–40 (*See also* pedagogy and instruction); B.Ed. degree, 169, 173; courses, 3, 11, 18, 48–49, 51, 57, 65, 69, 84, 98–99, 132, 139–40, 142–43, 185, 193, 215–16, 219; Qualified Teacher Status (England), 169, 181 n.17

PUPA (Proof Understanding, Production and Appreciation) Project, 185, 210 n.3

Pythagoras (c.570–c.495 B.C.), 1, 12

quadratic residue, 166–167

quasi-induction, 195–203, 206, 209, 228 (*See also* MI; proof scheme); cognitive root of MI, 201; process pattern generalization, 195–201, 206, 229; repeated reasoning, 197, 209–10, 229

quotient (whole number), partitive disposition (magnitude of resultant parts), 20–21, 35–36, 37; quotitive disposition (closest multiple of divisor to dividend), 20, 21, 26–28, 35–36, 37; as sum of quotient and remainder, 28–29, 30; in terms of divisor, 22, 23–24; in terms of rational quotient, 20–21, 30; in terms of remainder, 28, 34; in terms of the "result" or "answer," 29, 34; in terms of unity, 21, 22; various meanings of, 34–35. *See also* division with remainder; remainder

reasoning, 9, 44, 73–74, 139, 140, 197 (*See also* conjectures; generalization; MI; proof; proof scheme; quasi-induction); approximate, 223; deductive, 222; inductive (inference), 135, 159, 191; norms for argumentation, 140–42; valid arguments, 222

recursion, 190–91, 195, 203, 207

reflection, 220; reflecting on actions and processes, 43, 46, 78–79, 178, 224–25

reflective discourse, 11, 125, 126, 128

remainder, as sum of quotient and remainder, 24; in terms of fractions and decimals, 20–21, 24, 25–26, 27, 30, 33–34, 35, 36, 37–38; in terms of quotient, 21–22, 34; in terms of "what's left (over)," 22, 27, 219. *See also* division with remainder; quotient

representation, 9, 11, 28, 79, 106, 121, 124–25, 128, 135, 218; and APOS theory, 106–7, 113; base, 26; decimal (form), 22, 24–28, 33, 35, 38, 50, 56–58, 61, 64, 65, 78–80, 91, 92, 102, 103, 111, 112, 133, 134, 136, 220; invariance, 102, 160–61; multiple (meanings), 38, 98, 102–4, 125, 218, 220; prime factor, 42, 109, 111, 134, 136 (*See also* factors, prime; prime numbers); semiotic, 100; symbolic, 7, 10, 17, 120–21, 128, 226; unexpected and unfamiliar, 100–1, 228; verbal, 128

rigor, 83, 93–94, 186, 226; proof 11, 186; thinking 80

RUMEC (Research in Undergraduate Mathematics Education Community), 81 n.1

schema theory. *See* APOS theory; conceptual field; proof scheme

scheme, 6, 210 n.5. *See also* proof scheme

Social Sciences and Humanities Research Council of Canada, 39 n.1

teaching. *See* pedagogy and instruction
teaching experiment, 188–89
theorem-in-action, 17, 199. *See also* quasi-induction
Tower of Hanoi problem, 190

understanding, 207, 227 (*See also* knowledge; learning); abstract, 5–6, 8, 12; arithmetic, 32–33; conceptual (relational, structural), 1, 2, 21, 194, 227–28; conceptual, developing, 124; conceptual, limitations in, 11, 123; conceptual, of divisibility, 31, 58; conceptual, of integers, 1, 4, 8, 11, 12; conceptual, procedural in absence of or prior to, 80, 113, 116, 227–28; discursive, 126; novel, unfamiliar, and abstract problem contexts, 18, 19, 20, 21–24, 36, 41–42; procedural, 8, 80, 113, 116, 118, 194; PUFM (Profound Understanding of Fundamental Mathematics), 11, 80; "success to understanding," 45–46, 227–28; variable, 123, 128; variable, as placeholder, 123
University of Cambridge, 169, 173

visual 89, components and features, 99, 149, 153; (de)composition, 147–48; images of multiplication and division, 89, 218; metaphor, 215; patterns, 148; representations, 103–4

whole number rule, 33, 35
Wilson, J. (1741–1793), 165, 174, 179
Wilson's theorem, 165, 174–75, 176, 179

zero, condition for, divisibility ("divides evenly"), 16, 17, 36, 91–91, 217, 226 (*See also* divisibility); condition for, multiples, 62; condition for, non-existent primitive roots, 171; condition for, polynomial roots, 105; condition for, semigroups, 215; division by, 15; fractional component, 16, 17, 35; placeholder, 57; remainder, 16, 17, 35

About the Contributors

ANNE BROWN is an Assistant Professor in the Department of Mathematical Sciences at Indiana University, South Bend. She conducts research on the learning of collegiate mathematics, emphasizing in particular the ways in which mathematical knowledge for teaching develops. Her recent work focuses on the learning of fundamental concepts in abstract algebra, number theory, and set theory.

STEPHEN R. CAMPBELL is an Assistant Professor in the Department of Education at the University of California, Irvine. He is mainly interested in how the cognitive history of philosophy and mathematics can inform mathematics education. His work in learning and teaching number theory emphasizes the role of developing a conceptual understanding of arithmetic as a gateway to algebra. He is currently developing new methodologies for research in mathematical cognition using computer-based learning environments.

LAURIE D. EDWARDS is Associate Professor of Education at St. Mary's College of California. She has developed and researched computer-based microworlds in geometry and elementary mathematics, and has investigated informal proof and pre-proof thinking. She is interested in number theory as a domain for investigating problem solving and reasoning, and in embodied cognition as the foundation for mathematical understanding.

PIER LUIGI FERRARI is an Associate Professor working in the area of Mathematics Education in the Department of Science and Advanced Technology at the University of East Piedmont, Alessandria. His main interests concern the role of language in mathematics education. He is also engaged in developing new teaching methods for introductory mathematics courses at the undergraduate level. His work in learning number theory emphasizes the role of semantic control in learning processes.

GUERSHON HAREL is a Professor in the Mathematics Department at the University of California, San Diego. His research interests are in cognition and epistemology of mathematics and their application in mathematics curricula and teacher education. He has focused on students' conceptions of mathematical proof, the learning and teaching of linear algebra, and the development of proportional reasoning.

TIM ROWLAND is a Lecturer in Mathematics Education at the University of Cambridge, England, and an active mathematician in the field of Number Theory. His recent book *The Pragmatics of Mathematics Education* reflects a range of current research interests in mathematics, linguistics, epistemology, and education. Another research strand with colleagues in London, Durham, and York investigates the relationship between teachers, mathematics subject knowledge, and their classroom performance. He is an executive committee member of the British Society for Research into the Learning of Mathematics.

ANNIE SELDEN is a Professor of Mathematics at Tennessee Technological University and Coordinator of the Mathematical Association of America's Special Interest Group for Research in Undergraduate Mathematics Education. She is co-editor of the volumes *Research in Collegiate Mathematics Education* and associate editor for Teaching/Learning of MAA Online, for which she also writes (with John Selden) the ''Research Sampler'' column. Her current research interests include mathematical reasoning and problem solving, and students' understanding of logic, definition, proof, and validation.

JOHN SELDEN is a retired Professor of Mathematics who served as Department Head and Dean of Science at Bayero University in Nigeria before teaching at Tennessee Technological University. He has supervised nine mathematics Ph.D. dissertations in semi-groups. He has been a visiting scholar in mathematics education (with Annie Selden) at the University of California, Berkeley, San Diego State University, and Arizona State University. His current research interests include mathematical reasoning and problem solving, and students' understanding of logic, definition, proof, and validation.

ANNE R. TEPPO served as adjunct faculty in mathematics education at Montana State University, Bozeman, where she developed a one-semester mathematics course for preservice elementary education majors. Her areas of interest include working with math-anxious adults, developing innovative classroom materials, and research into the teaching and learning of algebra. She is the author of numerous articles and book chapters related to her work in these areas.

KAREN THOMAS is an associate professor in the Mathematics Department at the University of Wisconsin, Platteville. Her research interests focus on the learning and teaching of mathematics at the undergraduate level, including topics in calculus, linear algebra, and abstract algebra, and on the processes of development of pedagogical content knowledge in mathematics for prospective elementary and middle-level teachers.

GEORGIA TOLIAS is a Visiting Assistant Professor at DePaul University, Chicago. She holds a joint appointment with the Department of Mathematical Sciences and the Quantitative Reasoning Program. Her scholarly activity is in undergraduate mathematics education with an emphasis on the synthesis between a particular constructivist theory of learning and corresponding pedagogical techniques. Her current research study investigates students' understanding of the concept of percent as it appears in various contemporary quantitative contexts.

RINA ZAZKIS is a Professor of Mathematics Education at Simon Fraser University, Vancouver, British Columbia. Her research is in undergraduate mathematics education with a focus on mathematical content knowledge of preservice teachers and the ways in which this knowledge is acquired and modified. She has conducted research in undergraduate students' learning and understanding of group theory and set theory. Her recent studies on learning number theory have been supported by grants from the Social Sciences and Humanities Research Council of Canada.